Russell Thacher Trall

The Human Voice

It's Anatomy, Physiology, Pathology, Therapeutics, and Training

Russell Thacher Trall

The Human Voice
It's Anatomy, Physiology, Pathology, Therapeutics, and Training

ISBN/EAN: 9783744692601

Printed in Europe, USA, Canada, Australia, Japan

Cover: Foto ©berggeist007 / pixelio.de

More available books at **www.hansebooks.com**

THE HUMAN VOICE.

WORKS BY THE SAME AUTHOR.

Hydropathic Encyclopedia,	$4 40	Alcoholic Controversy,	50
The Hygienic Hand Book,	2 00	Water Cure for the Million,	30
Sexual Physiology,	2 00	True Healing Art,	30 or 50
Uterine Displacements,		Hydropathic Cook Book,	$1 50
Colored Plates,	5 00	Family Gymnasium,	1 50
Throat and Lungs,	25	Home Treat. Sex. Abuse,	50
Digestion and Dyspepsia,	1 00	The Bath—Its Uses,	25 or 50
Mother's Hygienic Hand-Book,	1 00	Hygeian Home Cook Book,	25
		Hygienic Catechism,	10

A Set of Six Anatomical and Physiological Plates, $20.

THE
HUMAN VOICE:

ITS ANATOMY, PHYSIOLOGY, PATHOLOGY, THERAPEUTICS, AND TRAINING;

WITH

RULES OF ORDER FOR LYCEUMS.

BY

R. T. TRALL, M.D.,

PRINCIPAL AND FOUNDER OF THE HYGEIO-THERAPEUTIC COLLEGE; PROFESSOR OF INSTITUTES OF MEDICINE, AND AUTHOR OF NUMEROUS WORKS.

NEW YORK:
S. R. WELLS & COMPANY.
737 BROADWAY.
1875.

COPYRIGHT, 1875, BY S. R. WELLS & CO.

PREFACE.

The object of this little work is to present, in a cheap and convenient form, the facts and principles applicable to the culture and uses of the Human Voice, which are only to be found scattered through several large volumes, and to furnish Lyceums and Debating Clubs with a concise Code of Rules and Usages for the regulation of their proceedings. It is not expected nor intended to supersede the more elaborate works on Elocution, which may be indispensable for the Orator and Teacher; but to furnish all who desire to read and speak well, and who must rely mainly on self-education, with a plain and intelligible guide in theory and practice.

R. T. T.

Florence Hights, N. J., 1875.

CONTENTS.

CHAPTER I.—ANATOMY OF THE VOICE.

Apparatus of Voice—The Thorax—Heart and Lungs—Ligaments of the Larynx—The **Larynx** Laterally—Muscles of the Larynx—Abdominal Muscles—Muscles of the Trunk—Muscles of the Trunk **Laterally**—Action of the Diaphragm—Range of the Human Voice—Bass and Tenor—Contralto and Soprano—Tone of Voice—Falsetto Voice, . 9

CHAPTER II.—PHYSIOLOGY OF THE VOICE.

Erectitude—Natural Spine—Vocal Cords—Pitch of the Tones—Volume of Voice—Character of Voice—Rationale of Respiration—Rationale of Sobbing and Laughter—Rationale of Speech—Vowel and Consonant Sounds — Whispering — **Sighing** — Ventriloquism — Speaking Automata—Rationale of Articulate Sounds—Distinctions of the Consonant Sounds, ... 18

CHAPTER III.—PATHOLOGY OF THE VOICE.

Causes of Defective Voice—Spinal Miscurvature—Natural and Deformed Chest—Positions in Study—Positions in Standing—Sleeping with the Mouth Open—Lisping—Stammering—Hoarseness—Aphonia—Nasal Tone—Vailed Tone—Explosive Vocalization, . - . 28

CHAPTER IV.—THERAPEUTICS OF THE VOICE.

Exercises to Improve Respiration—Walking—Slapping the Abdomen—Apparatus—Military Position—Rotating the Arms—Elbow Whirl—Chest Extension Exercises—Indian Club Exercises—Back-boards and Bands—Exercises with Weights—Directions for Lispers and Stammerers, ... 33

CONTENTS.

CHAPTER V.—TRAINING OF THE VOICE.

PAGE

Normal Positions—Declamation—Argument—Exhortation—Appeal—Controlling the Respiration—Full Breathing—Audible Breathing—Forcible Breathing—Sighing—Gasping—Panting—Management of the Voice—Regulation of Tones—Enunciation—Deportment, . . 41

CHAPTER VI.—EXERCISES ON THE ELEMENTARY SOUNDS.

Analysis of the Elementary Sounds—Analysis of the Sounds of the Letters—Exercises on the Vowel Sounds—Exercises on the Consonant Sounds—Exercises in Emphasis—Examples of Intonations—Examples of Waves or Circumflexes, ... 47

CHAPTER VII.—SELECTIONS FOR PRACTICE.

To Range—Glory—Cato's Soliloquy—Our Honored Dead—Darkness—Curtain Lecture—Immortality—Advantages of Adversity—Morning—The Dilemma—Deity—The Death of Hamilton—The Stars—Public Virtue—Criticism—The Revolutionary Alarm—Sheridan's Ride—The Raven—The Bells—Christmas—The Tomahawk submissive to Eloquence, ... 56

THE HUMAN VOICE.

CHAPTER I.

ANATOMY OF THE VOICE.

The special apparatus of the voice is the larynx, an arrangement of ligaments and muscles at the upper part of the windpipe (trachea). The quality of voice depends on the tension and approximation of the vocal cords; its depth or fullness depends on the capacity of the chest, and its power on the associated action of all the respiratory muscles. A brief exposition, therefore, of the structure of the vocal and respiratory apparatus seems to be necessary as a basis for the intelligent training and proper exercise of the organs of music and speech.

The foundation for a normal voice as well as for bodily and mental vigor, and, indeed, for good health, is a well-developed thorax, or framework of the chest.

This is constituted of the sternum, or breast-bone, in front, and the twelve pairs of ribs on the sides. The ribs are articulated behind with the twelve dorsal vertebræ of the spinal column. The trachea commences opposite the fifth cervical vertebræ, and extends to the third dorsal, where it divides into the right and left bronchi, which pass to the right and left lung, and are subdivided and ramified throughout the substance of the lungs. The trachea and bronchial tubes are everywhere lined with a mucous membrane, as is the mouth and larynx. Two-thirds of the anterior cylinder of the

Fig. 1.

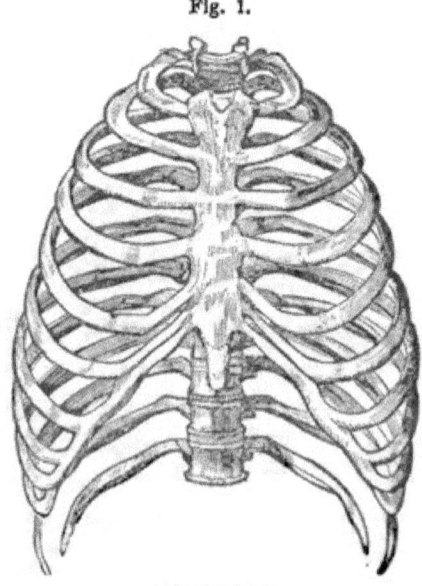

THE THORAX.

An anterior view of the thorax is represented in Fig. 1. 1. The manubrium. 2. Body. 3. Ensiform cartilage. 4. First dorsal vertebra. 5. Last dorsal vertebra. 6. First rib. 7. Head of first rib. 8. Its neck. 9. Its tubercle. 10. Seventh rib. 11. Costal cartilages of the ribs. 12. Last two false ribs. 13. The groove along the lower border of each rib.

trachea are composed of fifteen to twenty cartilaginous rings, which are conducive to the vibrations of air in making trilling sounds.

The thyroid gland (sometimes the seat of goitre, or bronchocele,) is situated upon the trachea above the sternum; it is divided into two lobes, one of which is placed on each side of the trachea.

The lungs occupy the cavity of the chest on each side of the heart. They are conical in shape, tapering above, where they extend beyond the level of the first rib, and broad and concave below, where they rest on the convex surface of the diaphragm. The root or upper portion of

Fig. 2.

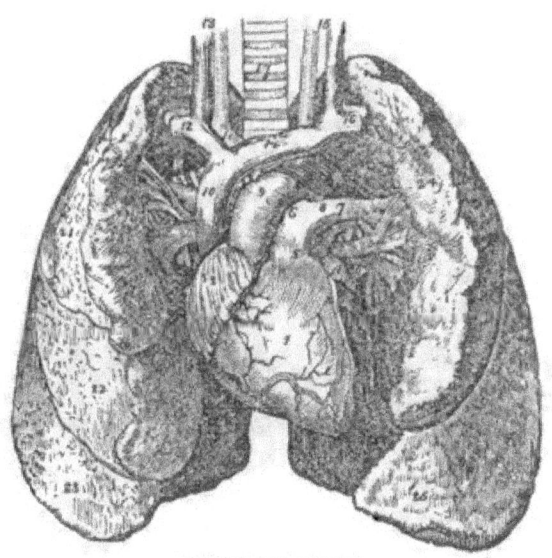

HEART AND LUNGS.

Fig. 2 represents the anterior aspect of the anatomy of the heart and lungs. 1. Right ventricle; the vessels to the left of the number are the middle coronary artery and veins. 2. Left ventricle. 3. Right auricle. 4. Left auricle. 5. Pulmonary artery. 6. Right pulmonary artery. 7. Left pulmonary artery. 8. Remains of the ductus arteriosus. 9. Aortic arch. 10. Superior cava. 11. Arteria innominata; in front of it is the right vena innominata. 12. Right subclavian vein; behind it is its corresponding artery. 13. Right common carotid artery and vein. 14. Left vena innominata. 15. Left carotid artery and vein. 16. Left subclavian artery and vein. 17. Trachea. 18. Right bronchus. 19. Left bronchus. 20, 20. Pulmonary veins; 18, 20, from the root of the right lung; and 7, 19, 20, the root of the left. 21. Upper lobe of the right lung. 22. Its middle lobe. 23. Its inferior lobe. 24 Superior lobe of left lung. 25. Its lower lobe.

each lung, which retains the organ in position, comprises the pulmonary artery and veins, the bronchial tubes, the bronchial vessels, and the pulmonary plexuses of nerves.

ANATOMY OF THE VOICE.

The minute anatomy of the larynx is shown in figs. 3 and 4.

Fig. 3.

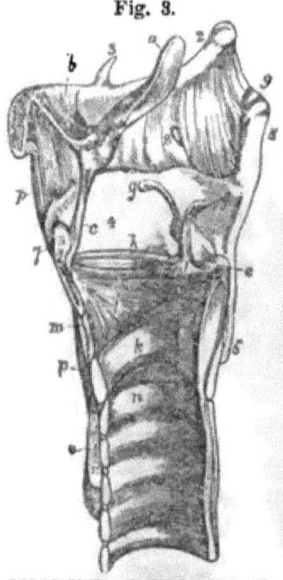

LIGAMENTS OF THE LARYNX.

Fig. 3 is a vertical section of the larynx, showing its ligaments. 1. Body of the os hyoides. 2. Its great cornu. 3. Its lesser cornu. 4. The ala of the thyroid. 5. The superior cornu. 6. Its inferior cornu. 7. Promum Adami. 8, 8. Thyro-hyoidean membrane; the opening near the posterior numeral transmits the superior laryngeal nerve and artery. 9. Thyro-hyoidean ligament. *a.* Epiglottis. *b.* Hypo-epiglottic ligament. *c.* Thyro-epiglottic. *d.* Arytenoid cartilage. *e.* Outer angle of its base. *f.* Corniculum laryngis. *g.* Cuneiform cartilage. *h.* Superior thyro-arytenoid ligament. *i.* Chorda vocalis, or inferior thyro-arytenoid; the elliptical space between the two thyro-arytenoid; is the laryngeal ventricle. *k.* Cricoid cartilage. *l.* Lateral portion of the crico-thyroidean membrane. *m.* Its central portion. *n.* Upper ring of the trachea, which is received within the ring of the cricoid cartilage. *o.* Section of the isthmus of the thyroid gland. *p, p.* The levator of the glandulæ thyroideæ.

Fig. 4.

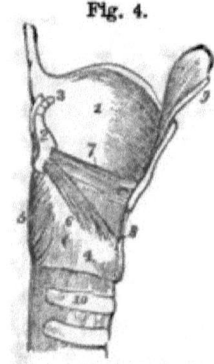

THE LARYNX LATE-RALLY.

Fig. 4 is a side view of the larynx, one ala of the thyroid cartilage being removed. 1. Remaining ala. 2. One of the arytenoid cartilages. 3. One of the cornicula laryngis. 4. Cricoid cartilage. 5. Posterior crico-arytenoid muscle. 6. Crico-arytenoidens lateralis. 7. Thyro-arytenoideus. 8. Crico-thyroidean membrane. 9. One half of the epiglottis. 10. Upper part of the trachea.

The following description of the laryngeal structures is copied from the "Hydropathic Encylopædia :

"The cartilages are : 1. *Thyroid* (shield-like), which consists of two lateral portions (*alæ*) meeting at an angle in front, and forming the projecting part of the throat, called *pomum Adami* (Adam's apple). Each ala forms a rounded border posteriorly, which ter-

ANATOMY OF THE VOICE.

minates above in a *superior cornu*, and below in an *inferior cornu*. 2. *Cricoid* (like a ring), a circular ring, narrow in front and broad behind, where it has two rounded surfaces, which articulate with the arytenoid cartilages. The œsophagus is attached to a vertical ridge on its posterior surface. 3. Two *arytenoid* (pitcher-like); triangular in form, and broad and thick below, where they articulate with the upper border of the cricoid; above they are pointed and prolonged by two small pyriform cartilages, called *cornicula laryngis*, which form part of the lateral wall of the larynx, and afford attachment to the chorda vocalis and several of the articulating muscles. 4. Two *cuneiform ;* small cylinders, about seven lines in length, and enlarged at each extremity; they are attached by the lower end to the arytenoid, and their upper extremity forms a prominence on the border of the arytenoepiglottidean fold of membrane; they are occasionally wanting. 5. *Epiglottis ;* shaped like a cordate leaf, and situated immediately in front of the opening of the larynx, which it closes when the larynx is drawn up beneath the base of the tongue, as in the act of swallowing. The laryngeal cartilages ossify more or less in old age, particularly in the male.

"The *ligaments* are : 1. Three *thyro-hyoidean*, which connect the thyroid cartilage with the os hyoides. 2. Two *capsular crico-thyroid*, which articulate the thyroid with the cricoid, and with their synovial membranes from the articulation between the inferior cornu and sides of the cricoid. 3. The *crico-thyroidean* membrane, a fan-shaped layer of elastic tissue, attached by its apex to the lower border of the thyroid, and by its expanded margin to the upper border of the cricoid and base of the arytenoid ; above it is continuous with the lower margin of the chorda vocalis. 4. Two *capsular crico-arytenoid*, which connect those cartilages. 5. Two *superior thyro-arytenoid*, thin bands between the receding angle of the thyroid and the anterior inner border of **each arytenoid**; the lower border constituting the upper boundary of the ventricle of the larynx. 6. Two *inferior thyro-arytenoid*, the *chordæ vocales*, which are thicker than the superior, and, like them, composed of elastic tissue. Each ligament, or vocal chord, is attached in front to the receding angle of the thyroid, and behind to the anterior angle of the base of the arytenoid. The inferior border of the chorda vocalis is continuous with the lateral expansion of the crico-thyroid ligament. The superior border forms the lower boundary **of the ventricle of the larynx.** The space between the two chordæ vocales is the *glottis* or *rima glottidis*. 7. Three *glosso-epiglottic*, folds of mucous membrane connecting the anterior surface of the epiglottis with the root of the tongue. 8. The *hyo-epiglottic*, an elastic band connecting the anterior aspect of the epiglottis with the hyoid bone. 9. The *thryo-epiglottic*, a slender elastic **slip** embracing the apex of the epiglottis, and inserted into the thyroid above the chordæ vocalæs.

" The *muscles* are eight in number: five larger ones of the chordæ **vocales** and glottis, and three smaller of the epiglottis. The **origin,** insertion, and use of each is expressed by its name. They are the *crico-thyroid, posterior* and *lateral crico-arytenoid, thyro-arytenoid, arytenoid thyro-epiglottic,* and *superior* and *inferior arytenoepiglottic.* The posterior crico-arytenoid **opens the** glottis; **the arytenoid** approximates the arytenoid cartilages **posteriorly,** and the crico-arytenoideus lateralis and thyroarytenoidei anteriorly; the latter **also close** the glottis mesially. The crico-thyroidei are tensors of the vocal chords,

ANATOMY OF THE VOICE. 15

and with the thyro-arytenoidei, regulate their position and vibrating length. The remaining muscles assist in regulating the tension of the vocal chords by varying the position of their cartilages.

"The *aperture* of the larynx is a triangular opening, broad in front and narrow behind; bounded in front by the epiglottis, behind by the arytenoid muscle, and on the sides by the folds of the mucous membrane. The cavity is divided into two parts by an oblong constriction produced by the prominence of the vocal chords; the part above the constriction is broad above and narrow below, and the part beneath is narrow above and broad below, while the space included by the constriction is a narrow, triangular fissure, the *glottis*, bounded on the sides by the chordæ vocales and inner surface of the arytenoid cartilages, and behind by the arytenoid muscle; it is nearly an inch in length, somewhat longer in the male than female. Immediately above the prominence caused by the chorda vocalis, and extending nearly its length on each side of the cavity of the larynx is the *ventricle of the larynx*, an elliptical fossa which serves to isolate the chord.

"The *mucous membrane* lines the entire cavity of the larynx, its prominences and depressions, and is continuous with that of the mouth and pharynx, which is prolonged through the trachea and bronchial tubes into the lungs. In the ventricles of the larynx the membrane forms a cæcal pouch, called *sacculus laryngis*, on the surface of which are the openings of numerous follicular glands, whose secretion lubricates the vocal chords."

The abdominal muscles are important parts of the respiratory machinery; comparing the lungs to a bellows, these muscles constitute the handles, and unless they are well developed and in vigorous condition, the voice be correspondingly feeble and imperfect. The relation of these muscles to the thorax directly, and to the lungs and vocal apparatus indirectly, is shown in figs. 5 and 6.

Fig. 5.

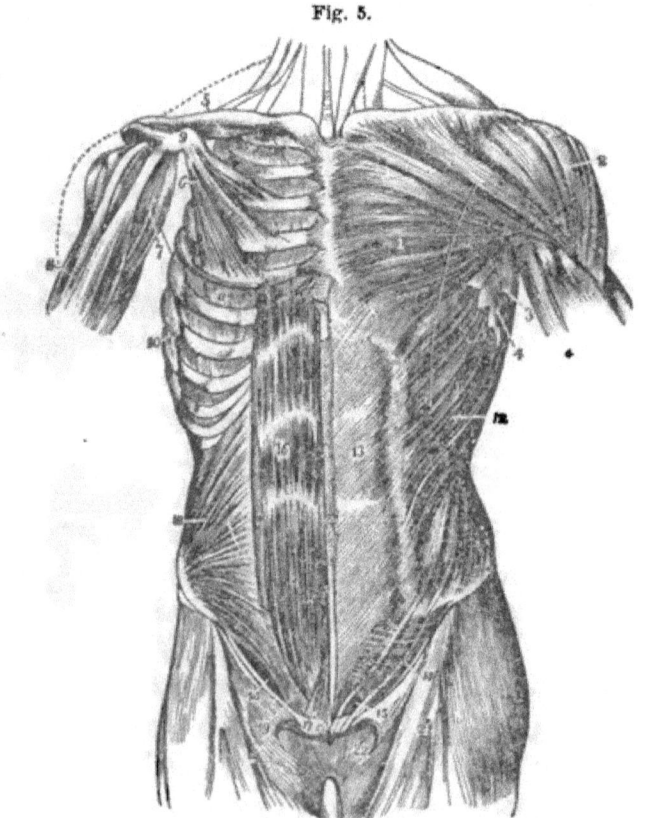

MUSCLES OF THE TRUNK.

In Fig. 5 are seen the muscles of the trunk anteriorly. The superficial layer is seen on the left side, and the deeper on the right. 1, Pectoralis major. 2. Deltoid. 3. Anterior border of the latissimus dorsi. 4. Serrations of the serratus magnus. 5. Subclavius of the right side. 6. Pectoralis minor. 7. Coracho-brachialis. 8. Upper part of the biceps, showing its two heads. 9. Coracoid process of the scapula. 10. Serratus magnus of the right side. 11. External intercostal. 12. External oblique. 13. Its aponeurosis; the median line to the right of this number is the linea alba; the flexuous line to the left is the linea semilunaris; the transverse lines above and below the number are the lineæ transversæ. 14. Poupart's ligament. 15. External abdominal ring; the margin above is called the

ANATOMY OF THE VOICE.

superior or *internal* pillar; the margin below the *inferior* or *external* pillar; the curved intercolumnar fibres are seen proceeding upward from Poupart's ligament to strengthen the ring. The numbers 14 and 15 are situated upon the fascia lata of the thigh; the opening to the right of 15 is called *saphenous*. 16. Rectus of the right side. 17. Pyramidalis. 18. Internal oblique. 19. The common tendon of the internal oblique and transversalis descending behind Poupart's ligament to the pectineal line. 20. The arch formed between the lower curved border of the internal oblique and Poupart's ligament beneath which the spermatic cord passes, and hernia occurs.

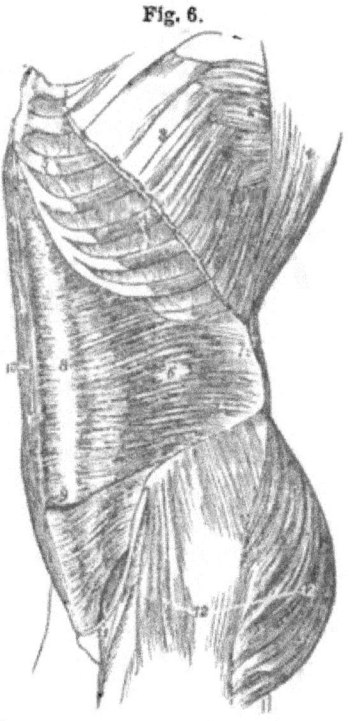

Fig. 6.

Fig. 6 is a side view of the muscles of the trunk. 1. Costal region of the latissimus dorsi. 2. Serratus magnus. 3. Upper part of external oblique. 4. Two external intercostals. 5. Two internal intercostals. 6. Transversalis. 7. Its posterior aponeurosis. 8. Its anterior. 9. Lower part of the left rectus. 10. Right rectus. 11. The arched opening where the spermatic cord passes and hernia takes place. 12. The gluteus maximus, and medius, and tensor vaginæ femoris muscles invested by fascia lata.

The oblique muscles flex the thorax on the pelvis; either acting singly, twists the body to one side. Either transversalis muscle by contracting diminishes the size of the abdomen, and both acting together constrict its general cavity. The recti muscles, and the pyramidalis pull the thorax forward when acting together. MUSCLES OF THE TRUNK LATERALLY.

All of the abdominal muscles are auxiliary to respiration, and as they constitute the chief forces in expelling the air from the lungs, their relation to voice is obvious. As respiratory muscles they are aided by the muscles of the loins and back; the united action of all these muscles compresses the abdomen in all directions, as may be noticed in prolonged coughing or severe vomiting.

CHAPTER II.

PHYSIOLOGY OF THE VOICE.

Fig. 7.

NATURAL SPINE.

PHYSICAL uprightness is as important for a public speaker as moral rectitude is for a private citizen. Other things being equal, every person will have a power to please and persuade, influence and direct the minds of others, through the media of speech and music, measurable generally by the integrity of the whole bodily organization, and especially by the erectitude of the spinal column (Fig. 7), without which the extensive and complicated machinery of respiration and vocalization cannot act harmoniously.

The vocal apparatus has been compared to a stringed, tubular, and reeded instrument, as the violin, flute, and clarionet; it has many properties in common with each, and, indeed, with all musical instruments; yet it differs in many respects from either. No mechanical contrivance can rival the variety and delicacy of action of the living structure,

PHYSIOLOGY OF THE VOICE.

hence the human voice must ever be incomparably superior, as a musical instrument, to all human inventions. A good reader, a good speaker, or a good singer never fails to attract the multitudes.

The lower vocal cords are chiefly instrumental in the production of *sound*. If the upper cords are removed, voice continues, but is rendered feeble; if the lower cords are destroyed, voice is entirely lost.

The *tones of voice* depend on the varying tension of the vocal cords. In producing tones, the ligaments of opposite sides are brought into approaching parallelism with each other, by the approximation of the points of the arytenoid cartilages; in the intervals they are again separated, and the opening between them, termed *rima glottidis*, assumes the form of the letter V, as represented in Fig. 8.

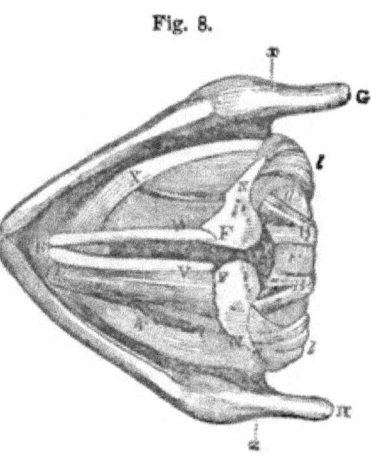

Fig. 8.

Fig. 8 exhibits the vocal ligaments as seen superiorly. G, E, H. Thyroid cartilage. N, F. Arytenoid cartilages. S, V, S, V. Vocal cords or ligaments. N, X. Crico-arytenoideus lateralis. V, k, f. Right thyro-arytenoideus. N, l, N, l. Crico-arytenoidei postici. B, B. Cricoarytenoid ligament.

LARYNX FROM ABOVE.

The muscles which stretch or relax the vocal ligaments, are alone directly concerned in the voice; the muscles which open and close the glottis, regulate the amount of the air inspired and expired.

The *pitch of the tones* is regulated by the tension of the vocal cords; its *volume* or intensity depends on the capacity of the lungs, length of the trachea, flexibility of the vocal cords, and the force with which the air is expelled from the lungs. The *character* of the voice is dependent on the confirmation of the pharynx, mouth, and nasal cavities. In the male the larynx is more prominent and the vocal cords are longer than in the female, in the proportion of three to two, which renders the voice in most cases an octave lower.

The free play of the diaphragm is an important factor in the volume of voice. To understand this matter fully it must be recollected that the movements of the respiratory apparatus are partly voluntary, for the purposes of being subservient to voice and speech, and partly involuntary, for the purposes of aerating the blood. The lungs themselves are entirely passive in respiration. When the walls of the chest are drawn asunder, and the thorax dilated by the action of the respiratory muscles, the atmospheric air rushes into the air-cells, distending them in proportion to the dilatation of the thorax, and keeping the surface of the lungs accurately in contact with the walls of the chest in all their movements. But if air be admitted into the cavity of the pleura, outside of the lungs, as by a penetrating wound, the lungs cannot be fully distended by inspiration, but will remain partially collapsed, although the thorax expands, for the reason that the pressure from without balances that within the air-cells. Fig. 9 illustrates the action of the diaphragm in respiration.

The diaphragm, by extending the ribs and pressing down the abdominal viscera, is the principal agent in inspiration. In a deep inspiration, the little muscles between the ribs (intercostals) assist in the expansion of the chest by spreading the ribs, aided also to some extent by the muscles of the thorax generally. Expiration,

PHYSIOLOGY OF THE VOICE.

as already stated, is mainly accomplished by the contraction of the abdominal muscles, which, by drawing down the ribs and compressing the viscera up against the relaxed diaphragm, diminish the cavity of the thorax from above.

Says Marshall (*Outlines of Physiology*): "The human vocal apparatus is analogous to a wind instrument with a double membranous tongue, the bronchi and trachea representing the windtube, the vocal cords the double

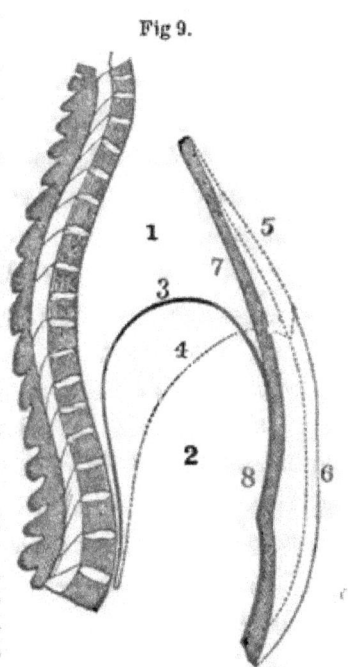

Fig. 9 is a side view of the chest and abdomen in respiration. 1. Cavity of the chest. 2. Cavity of the abdomen. 3. Line of direction for the diaphragm when relaxed in expiration. 4. Line of direction when contracted in inspiration. 5, 6. Position of the front walls of the chest and abdomen in inspiration. 7, 8. Their position in expiration.

ACTION OF THE DIAPHRAGM.

membranous tongue, and the parts above the glottis the attached tube. For the production of vocal sounds, even the feeblest, more air must pass through the glottis than in respiration; and this current of air must undergo penidic interruptions in its passage through that fissure. The vocal cords, moreover, are made more or less tense, and are approximated so as to be parallel to each other, and the fissure of the glottis is converted into a fine chink-like opening. The escape of the air propelled upward through the trachea being thus retarded, the margins of the vocal cords are forced upward, and slightly separated from each other; the elasticity of the cords is now called into play, so that they counteract the force of the impulse communicated to them, and, by a downward recoiling movement, again narrow the glottis. In this manner, the oscillations into which the vocal cords are thrown by the escape of the air driven from the trachea, or wind-tube, are communicated to the less tense air above the glottis, and throw this into vibrations. By means of the laryngeal ventricles, or sacs, placed above the vocal cords, these latter are kept free, so that their vibrations are easily accomplished. It has also been supposed by some, that the superior vocal cords maintain the strength and quality of the sounds, by entering into simultaneous and synchronous vibrations. This is contrary to Señor Garcia's observations with the laryngoscope; but he found that, in elevation of the pitch of the voice, whether natural or falsetto, the superior vocal cords approached each other, so as to narrow the part of the vocal tube above the glottis."

The ordinary range of the human voice, from the lowest male to the highest female voice, is nearly 4 octaves.

Fig. 10. E.
Fig. 11. C.
Fig. 12. F.
Fig. 13. A.

The lowest note, E (Fig. 10), is caused by 80 vibrations per second, and the highest note, C (Fig. 11), by 1,024 vibrations per second. But in exceptional cases, the range may be nearly 5½ octaves, the lowest note, F (Fig. 12), being caused by 42, and the highest note, A (Fig. 13), by 1,708 vibrations.

In ordinary speech, the range of voice is usually about half an octave; but in singing, the compass of the voice in the same individual generally extends over 2 octaves. In rare cases it has extended over 3½ octaves. It has been calculated that no less than 240 different states of tension of the vocal cords are producible at will, each degree of tension modifying the sound of the note in singing, or of the tone in speaking, and all this in a voice of ordinary range. Celebrated singers can produce a still greater number of intermediate tones. " Madame Mara," says Marshall, " could effect as many as 2,000 changes."

The bass and tenor varieties of voice are characteristic of the male, and the contralto and the soprano, otherwise known as second treble and first treble, of the female sex. The subdivision of voice called baritone, is intermediate between the tenor and bass, and the mezzo-soprano is intermediate between the soprano and contralto. The lowest note of the contralto is about an octave higher than the lowest note of the bass voice; and the highest soprano about an octave higher than the highest tenor.

The personal quality or peculiar tone of voice is due to the general confirmation of the air-passages; but in both sexes, more especially in the male, two series of notes can be produced, which have been distinguished into the true or chest notes, and the falsetto or head notes. The chest notes are called those of the natural voice, and are fuller, stronger, and more resonant, and are the lower notes of the voice; the falsetto notes are softer, less clear, and have a humming sound resembling the harmonic notes of strings. The middle notes of the scale can be produced by either the chest or the head voice. Some persons can speak or sing with either voice so well marked as to seem to be endowed with two distinct voices. Various theories have been advanced to account for the falsetto voice; but the observations of Garcia seem to prove that, during the production of the falsetto notes, the glottis is more elongated and widened, and that only the edges of the vocal cords are approximated, thus offering little resistance to the air, whilst, in the natural or chest voice, a certain depth of the surface of each cord is made to approach the other, and to undergo vibrations.

In certain strong mental emotions, the muscles of the voice act spasmodically, as in sobbing and laughter, and sometimes closing the glottis entirely for a longer or shorter time, as in some convulsive diseases.

Speech is the utterance of articulate sounds. The voice or vowel sounds are made with a nearly fixed position of the vocal organs; but as those sounds are modified by the action of the tongue, lips, etc., they are called articulate or consonant sounds. The vowel sounds are specially expressive of the feelings, while the consonant sounds are specially related to thought. This is why the language of music is so largely constituted of vowel sounds, the difference between music and speech consisting simply in the prolongation of the vowel sounds. As the language of all animals expresses much more of the affectional than of the intellectual mind, they have correspondingly little occasion for consonant sounds.

As vocalization depends on laryngeal vibrations, in whispering, vowels are articulated simply by the action of the mouth and fauces, all sound being produced above the larynx. Sighing is another example of sound produced above the larynx; if the vocal cords are called into vibratory action, the sigh becomes a groan. Most of the letters of the alphabet can be articulated with very little laryngeal action during inspiration.

Many sounds, as of smacking, clicking, kissing, and whistling, are generated in the mouth, and produced independently of laryngeal action.

Ventriloquism consists essentially in the imitation of

peculiar sounds. Its rationale is not well understood by physiologists. Magendie **supposed it to be** produced in the larynx. Some have **thought it was** produced simply by articulating while drawing in the breath. According to Muller, the sounds peculiar to ventriloquism **may be** made, after taking a deep inspiration, so as to occasion the protrusion of the abdominal viscera by the descent of the diaphragm, and maintaining the diaphragm in its depressed or contracted condition, by speaking during a very slow **expiration,** performed only by the lateral parieties of the chest, through a very narrow glottis.

Speaking automata have only succeeded in imitating the separate sounds **of the voice; they can never combine** them successfully so as to imitate the language **of** the vital organism.

The following lucid explanation of the various vowel and consonant sounds is copied from " Marshall's Physiology : "

"Articulate sounds are divided into vowels **and con-sonants.** The true *vowels*, or *open sounds*, **as** they are called, are generated in the larynx. They are merely uninterrupted vocal tones, variously modified in their outward passage, by alterations in the shape of the parts of the oral cavity through which they pass; thus, in uttering the pure vowel sounds, ā, ă, e, o, u, pronounced respectively as in the words far, fate, ell, old, and in French words containing the u, one and the same sound produced by the vibrations of the vocal cords is converted **into five** different sounds, by changes in the position of the tongue, and by the gradual prolongation **of the cavity** of the mouth, by means of the lips; the most natural of these vowel **sounds, or** the one most easily uttered, is the broad ā. In the same manner the *diphthong* sound, *i, ei, eu,* and the sounds of *y* and *w*, at the beginning of words are vocal tones, modified by further changes in the shape and form of the mouth.

"*Consonants*, or *shut sounds*, are entirely formed in the parts above the larynx, and are so named, because most, if not all of them, can only be sounded *consonantly*, that is, with another sound or vowel. They require, for their production, a shutting or valve-like action to take place, either between the lips, as in pronouncing the letters *b*, *p*, and *m;* or between the upper teeth and lower lip, as in the case of *f* and *v;* or between the tongue and the palate, as *d*, *g* hard, *c* hard, *k*, *q*, *t*, *r*, *l*, and *n*, or between the tongue and the teeth, as in the production of hissing sounds, such as *c* soft, or *s* and *z*. The *compound articulate sounds*, as *j*, or *g* soft, *ch* soft, *ch* guttural, *ph*, *sh*, *th*, *ng* and *x*, are produced by modifications, or combinations of some of the other pure consonant sounds. The aspirate *h* is produced by an increased expiratory effort, made with the mouth open, before a vowel or other sound.

"Those consonants which are produced by, or connected with, a sudden stoppage of the breath at a certain point, the opening leading from the pharynx to the nose being quite closed, and all the respired air passing through the mouth, are called *explosive* consonants. They are of two kinds: the simple explosive consonants, *b*, *d*, and *g* hard; and the aspirate explosives, *p*, *t*, *k;* these, when uttered, are unaccompanied by a vocal sound, that is, they are attended with an intonation of the voice. Those consonants which can be produced without a complete stoppage of the breath previous to their utterance, are called *continuous;* some of these sounds are developed by the passage of the air, with a degree of friction through the

PHYSIOLOGY OF THE VOICE.

mouth; in this way the consonants *v, f, s,* and *z*, are produced by **expiration** through the nose only, as *ng, m,* and *n.* In uttering the letters *l* and *r*, the air escapes through the nose and mouth; in pronouncing the first of these, the air escapes at the sides of the tongue; in pronouncing the sound, the tongue undergoes a vibrating movement. All the continuous consonants can be pronounced with a vocal sound, except the aspirate *h;* and some of them can be pronounced either with or without vocal intonation. Consonants have also been **named according to the seat of their production; thus** *p* **is called a** *labial, t* a *palatal, n* a *nasal*, and the Gaelic *ch* a *guttural* **consonant; but this** classification is exceedingly artificial and incorrect; for the greater number of articulate sounds are the result of the conjoined action of the mouth, lips, palate, and upper part of the air-passage."

CHAPTER III.

PATHOLOGY OF THE VOICE.

Fig. 14.

SPINAL MISCURVATURE.

The most common causes of imperfect respiration and defective voice are distortions of the spinal column, and contracted chests. No person with either deformity can have a powerful voice, whatever may be its other qualities. Fig. 14, SPINAL MISCURVATURE, is a representation of a very common form of spinal distortion, in which nearly all of the abdominal viscera are more or less displaced, and the respiratory muscles unbalanced and undeveloped. By contrasting this figure with that of the natural spine in the preceding chapter (fig. 6), the disastrous consequences of a crooked spinal column may be realized at a glance.

A single glance at the bones of the chest (fig. 1), is sufficient to show the injurious effects on the respiratory system directly, and the **vocal** organs indirectly of every **thing that** interferes **in the** least with the full expansion **of the lungs in breathing;** and the relation of the diaphragm **to** respiration **(fig. 9), explains** the horrid consequences of **tight-lacing.** That this subject may be fully comprehended, let us place the normal development of this vital part of the human being in contrast with the abominably abnormal condition so common in the society of fashionable American women.

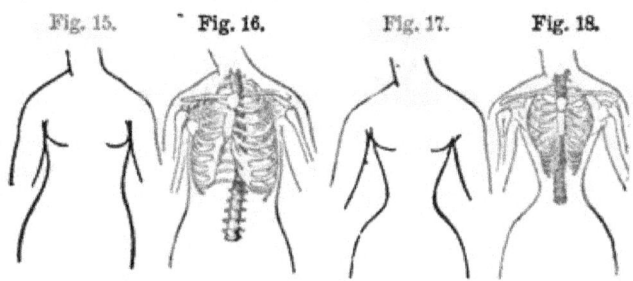

Fig. 15. Fig. 16. Fig. 17. Fig. 18.

NATURAL WAIST. NATURAL THORAX. CONTRACTED WAIST. FASHIONABLE WAIST.

A sufficient commentary on this fashionable folly and pernicious vice, so far as the immediate objects of this work are concerned, is the simple statement of the fact, that no female who deforms her body with tight-lacing ever becomes distinguished as a reader, speaker, or actor,

Fig. 19.

CORRECT POSITION IN STUDY.

Fig. 20.

MISPOSITION IN STUDY.

although a majority of them have attained distinction as chronic invalids and the mothers of feeble offspring.

All crooked bodily positions, by unbalancing the whole muscular system, enfeeble the breathing apparatus and impair the voice. Malpositions and spinal distortions are often acquired in the primary schools, because of the unanatomical construction of the miserable benches on which the suffering scholars are educated to "sit still" several hours each day. Figs. 19 and 20 illustrate this subject.

The malposition acquired in the sitting posture in

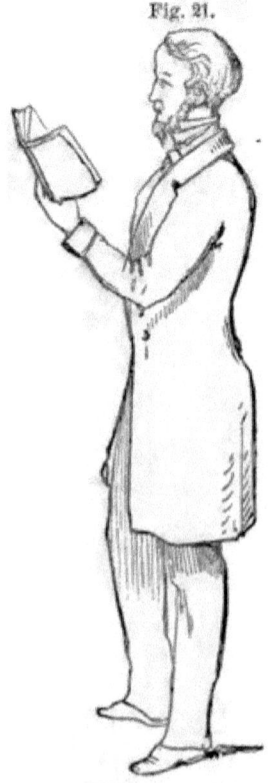

Fig. 21.

STANDING ERECT.

Fig. 22.

MALPOSITION IN STANDING.

childhood, is manifested in the standing **posture in adult life**, as represented in **fig. 22**, contrasted **with the perpendicular position, fig. 21.**

Although these deformities, which **are almost always acquired** in early life, can never be entirely overcome, much benefit may be derived from a persistent course of vocal culture, in connection with a proper system of gymnastic exercises; and if the laryngeal structures are favorably organized, **such** persons **may** become reputable speakers.

The habit of *sleeping with the mouth open* in early life, and especially in infancy, has **a very injurious effect on the breathing and** vocal organs; **and** not only **this, but it tends to distort the** jaw-bones and deform **the teeth.** Parents and nurses should be very careful to **check** this habit in its incipiency, or the damage may become irremediable. The **imperfections of speech** termed *lisping* and *stammering* are not attributable **to** organic defects, but **to** errors of action of the vocal apparatus. **In lisping the** tongue cleaves to the roof of the **mouth, or is pushed** against the upper teeth; stammering **is occasioned** by a spasmodic action of the glottis, tongue, **or** lips, which is **always** aggravated by any mental apprehension or embarrassment.

Hoarseness of **voice is usually** occasioned **by a swelling or congestion of the mucous membrane which lines the mouth, nose, trachea, or bronchial tubes. When the** laryngeal **portion of the mucous membrane is extremely** congested, **voice is entirely lost, as happens in some cases of** quinsy, diphtheria, **and croup, and in the later stage of** laryngeal consumption. **A chronic thickening of** the mucous membrane of **the laryngeal sacs or ventricles sometimes occasions permanent hoarseness, or complete loss of** voice. Paralysis of any one or more of the muscles of articulation may cause defect or loss of voice.

A common cause of defective voice, and sometimes of complete aphonia, is a want of association or co-operation in the action of the abdominal muscles and diaphragm in vocalization—a condition which may be occasioned by bodily malpositions, disease, or an improper use of the respiratory and vocal organs.

The *nasal tone* of voice is due to an approximation of the arches of the palate, more than to a closure of the nostrils.

The *vailed tone* of voice is occasioned by lowering the larynx so that it is covered by the entire pharynx, the base of the tongue being approximated to the palate, and the voice resounding in the upper part of the pharynx beneath the skull.

The *explosive voice*, which is due to the respired air being all passed out at the mouth, is always aggravated by a feeble co-operation of the abdominal muscles with the vocal effort. In this case the speaker becomes hoarse with any prolonged vocal effort. The explosive voice, though harsh and loud, is never heard at a great distance.

CHAPTER IV.

THERAPEUTICS OF THE VOICE.

PREMISING that, in all conditions of infirmity or disease affecting the voice, the general health is the first of all things to be attended to, this chapter will be devoted to such exercises and remedies as are specially applicable to defects of the vocal apparatus. In all cases it is important to harmonize as much as possible the action of all the muscles directly or indirectly concerned in respiration and voice; and just to the extent that this is accomplished will the disabilities or deformities be remedied, weak muscles and organs invigorated, obstructions removed, congestions reduced, and vocalization improved.

One of the best exercises is rapid walking over an uneven surface, or up and down stairs, keeping the mouth shut. The exercise should be commenced with moderation, and gradually increased in rapidity as can be borne without panting or difficulty of breathing. Those who are dyspeptic can improve the effect of this exercise by slapping the abdominal muscles as recommended in the author's work on "Digestion and Dyspepsia."

Among the "modern improvements" introduced into many health institutions, more or less useful for our purposes, are the health-lift, vibrator, dumb-bells, wands, rings, clubs, and other apparatus and machinery, each having special adaptation to some one or more of the many abnormal conditions prevalent. But for the benefit of those who are obliged to depend on self-treatment, a few illustrations, specially adapted to the respiratory and vocal apparatus, are copied from the author's "Illustrated Family Gymnasium," to which the reader is referred for a greater variety of illustrations. But in all exercises without apparatus the principle of bodily erectitude must be kept steadily in view or nothing will be gained. All bending of the body must be at the hip-points, and in lying, sitting, standing, walking, or running, playing, or working, no position must be maintained that bends the trunk of the body or in any manner restricts the play of the lungs, or compresses the abdominal viscera. The proper hint on this subject is afforded in the familiar calisthenic illustrations (Figs. 23 and 24).

Fig 23. ATTENTION. Fig. 24. MILITARY POSITION.

Keeping in mind the proper military "attention" under all circumstances, the circular exercise of the arms will be found an admirable one for bringing into gentle and equal action the whole respiratory system (Figs. 25 and 26).

This exercise is performed by extending the arms forward at right angles with the body, the palms of the hands being turned toward each other, and then rotating the arms alternately, then both together on the shoulder joint. Count one at each rotation, and turn the hands, during the movement, as far as possible both ways, so as to secure the rolling motion of arms and joints. After the movement has been performed half-a-dozen times in one direction, reverse it, and make as many movements

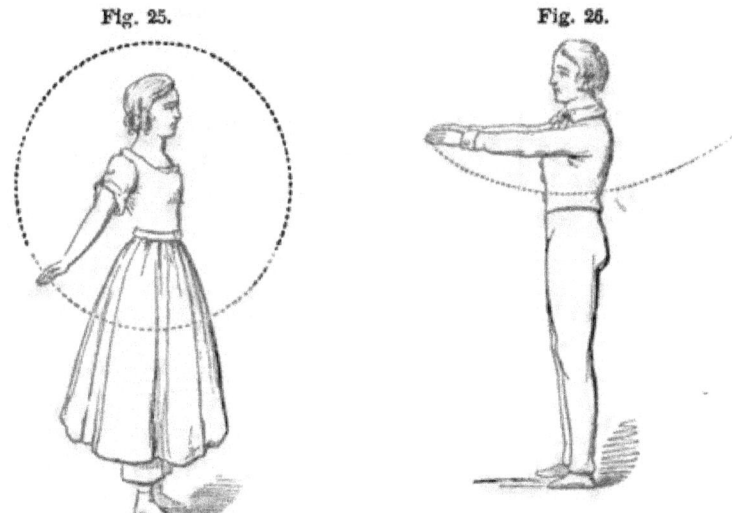

Figs. 25 and 26.—EXERCISE FOR THE WHOLE RESPIRATORY SYSTEM.

in the opposite direction; keep the palm of the hand down whenever the arm is raised.

The elbow whirl (fig. 27) may be performed as a variation of the above, and for very feeble persons, especially those troubled with shortness of breath, it is a good preparatory exercise. Place the elbows on the hips, and then swing the forearms in a circle.

The "circular" and "whirl" motions may be performed with increased effect while walking up hill or on an uneven surface.

The "lateral body swing" (fig. 28) is also an admirable preparatory exercise, and may vary the elbow whirl. This movement consists in bending the body from side to side, the arms being extended. It should be

ELBOW WHIRL.

performed very slowly at first, counting in a prolonged monotone to correspond with the bodily motions.

Fig. 28.

LATERAL MOVEMENT.

For those whose chests are contracted, who are round or stoop-shouldered, or who are predisposed to consumption, the "chest extension" exercise is especially to be recommended (figs. 29, 30, and 31).

These exercises comprehend several movements of the arms, all of which are intended to stretch the muscles and ligaments more especially of the upper part of the chest. Hold the arms at right angles with the body, and then throw the arms and hands backward and forward with considerable force, counting at each backward motion. Then from the same commencing position, vary

CHEST EXTENSION EXERCISES.

the exercise by striking the elbows together behind (fig. 31).

For the benefit of those who have the ordinary calisthenic apparatus, the following familiar illustrations are given:

The Indian club exercise is calculated to develop powerfully the muscles of the arms and chest. Figs. 32, 33, 34, and 35 show the principal positions so far as club exercises especially affect the respiratory system.

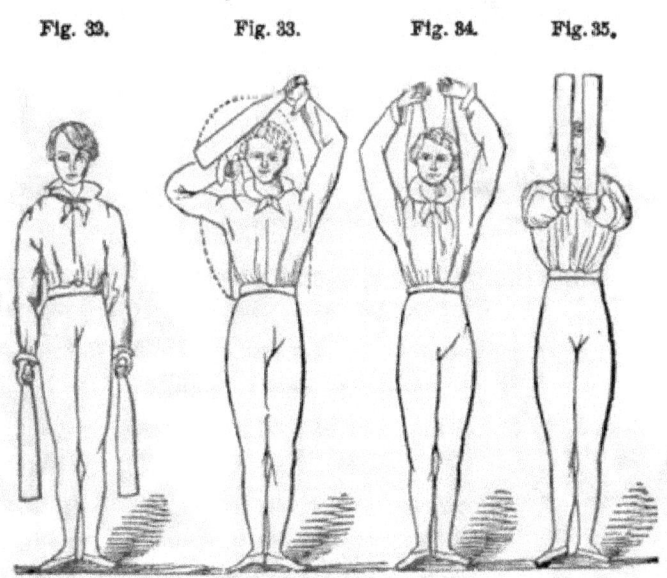

Fig. 32.   Fig. 33.   Fig. 34.   Fig. 35.

INDIAN CLUB EXERCISES.

Weights and dumb-bells may be employed to intensify the effect of any of the exercises which are usually performed without apparatus; and in a variety of such other ways as any one, understanding the object in view, can readily extemporize. Figs. 36 and 37 are examples.

Backboards and bands, which require no special explanation, help make a variety of useful apparatus. Figs. 38 and

39 represent some of the usual methods of exercising with them.

EXERCISES WITH WEIGHTS.

The impediments of speech termed *lisping* and *stammering*, can generally be remedied without difficulty by a persevering course of vocal training, and reasonable attention to hygienic conditions. The fact that those who lisp and stammer in speaking, usually articulate well enough in singing, suggests the proper remedial plan. They should aim to get entire mental control of the vocal apparatus by enunciating all of the elementary sounds of the language very slowly,

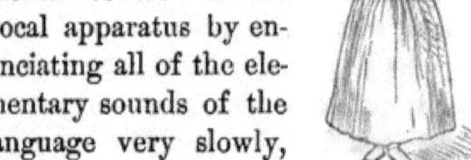

EXERCISES WITH BACKBOARDS.

deliberately, and distinctly, until the habit of convulsive action of the affected muscles is overcome. The stammerer should always speak with an expiring breath, and with the mouth well opened; a cure can generally be accomplished in a few months, sometimes in a few weeks. Indeed, a proper and persevering course of vocal gymnastics will almost certainly remedy the worst kind of stammering.

The first thing for the stammerer to do is to get complete control of his breathing apparatus. This can be done by means of the exercises mentioned in the succeeding chapter, especially those recommended by Professor Zachos, combined with the practice of slow, deep, full, and prolonged respirations. After this is accomplished, exercises on the vowel sounds, as explained hereafter, will be in order, constituting what M. Chevril, of France, who has acquired a reputation for the successful treatment of vocal impediments, terms the "gymnastics of articular phonation." When these vowel sounds are so thoroughly mastered that they can be distinctly enunciated forward and backward (thirty-two sounds) with a single expiration, and without any appreciable tendency to spasmodic action, the consonant sounds should be practiced on until all of them can be enunciated without the least inclination to stammer. Lastly, all of the elementary sounds of our language (forty-four), as explained in the ensuing chapter, should be practiced on until every sound is made without difficulty. Says M. Chevril: "The whole plan consists in gymnastically educating the organs of speech, the excellent results being due not so much to actual muscular work as to the precision with which the practice is carried out. The success depends on an effort of the will on the part of the patient to reproduce with the utmost precision a particular sound. The will of the teacher must take the place of the patient's wiL, as the latter is unable to regulate the movements dictated by it."

The principle above indicated may be readily comprehended when it is considered that hiccough, which is a spasmodic action of some of the respiratory muscles, can always be arrested instantly by a strong effort of the will properly directed. It is only necessary to fix the attention on some subject or object intensely; for example, the patient may determine to speak the word hiccough, *during* the next "attack," or paroxysm, and then watch intently for the first indication of it. If his attention is intense enough he will not hiccough again.

CHAPTER V.

TRAINING OF THE VOICE.

In all exercises having in view the improvement of the vocal apparatus, the first consideration as already stated, is a correct bodily position. It should be easy, unconstrained, and in all respects natural, allowing the freest play to every muscle concerned in respiration as well as vocalization. Figs. 40, 41, 42, and 43 represent some of the normal positions in public speaking.

PRESERVATION OF THE VOICE.

The rules for ensuring the durability and best working condition of the voice are few, simple, and mainly negative.

1. Be temperate in all things—and this means, avoid

Fig. 40.

DECLAMATION.

Fig. 41.

ARGUMENT.

gluttony and dissipation, and be moderate in all sensuous indulgences.

2. Do not make violent vocal efforts soon after a full meal; nor exert the voice at its highest pitch long at a

Fig. 42. EXHORTATION.   Fig. 43. APPEAL.

time. Never use the voice except very moderately when affected with hoarseness.

3. Butter, nuts, old cheese, sugar, candies, salted meats, acid liquors, ice-cream, very cold drinks, and very hot drinks, are especially injurious to the voice.

CONTROLLING THE RESPIRATION.

Among the essentials of good reading or speaking is a perfect command of the breath, so that all of the expired air can be used to the utmost advantage in vocalization. To acquire this condition:

1. Read or declaim in a *loud* whisper. This exercise is very fatiguing at first, and should be practiced but a few minutes at a time, until habit renders it easy.

2. Read or declaim in a low, strong key, passages which

require a firm and dignified enunciation, gradually proceeding to more spirited, and finally to the most impassioned sentences.

3. The following respiratory exercises, recommended by Prof. Zachos, are admirable for enabling the speaker to express the deeper emotions:

Full Breathing.—Stand in an erect position, with the arms akimbo, the hands resting on the hips; slowly draw in the breath until the chest is fully expanded; emit it with the utmost slowness.

Audible Breathing.—Draw in the breath as in full breathing, and expire it audibly, as in the prolonged sound of the letter k.

Forcible Breathing.—Fill the lungs, and then let out the breath suddenly and forcibly, in the manner of an abrupt and whispered cough.

Sighing.—Fill suddenly the lungs with a full breath, and emit as quickly as possible.

Gasping.—With a convulsive effort inflate the lungs; then send forth the breath more gently.

Panting.—Breath quickly and violently, making the emission of breath loud and forcible.

MANAGEMENT OF THE VOICE.

The proper management of the voice comprises due attention to tones, accent, emphasis, pronunciation, articulation, and pauses. The following rules should be observed:

1. Commence speaking a little *under* the ordinary pitch of voice.

2. The principal part of a discourse should be delivered in the ordinary pitch of voice; the exordium should be very deliberate and below the ordinary pitch, and the

peroration more impassioned and above the ordinary pitch.

THE REGULATION OF TONES.

Nothing is more awkward in public speaking than a misadaptation of tones to the occasion. They may be classified as follows:

1. The whisper, intended to be audible only to the nearest person.
2. The murmur, or low tone, adapted to close conversation.
3. The ordinary pitch, suited to general conversation.
4. The high or elevated pitch, adapted only to earnest argument or powerful appeals.
5. The extreme or highest pitch, appropriate only in the expression of violent passions.

ENUNCIATION.

Guard against the common fault of reading or speaking with the mouth insufficiently opened, or the teeth nearly closed. If this habit has been acquired, overcome it as speedily as possible. This may be done by reciting occasionally with a gag placed between the teeth; it may be made of card-board or a thin piece of wood. Commence with a gag half an inch wide, and gradually increase it to an inch and a half.

Be careful to articulate every syllable of every word. The general fault of readers, speakers, and especially singers, is in failing to articulate unaccented syllables. The rule of pronunciation is to regard every syllable as equally important, giving each its proper sound, and never slurring nor blending them together.

DEPORTMENT.

Under this head a few words on the countenance, manner, and gesture may be proper.

Nothing tends more to secure the sympathies of the audience than a quiet, self-possessed deportment. Never come before an audience, nor approach the speaker's desk in a hurried, bustling manner. Be deliberate and natural. *Be right, then act yourself.* Look over the audience, but do not stare *at* it. Avoid all awkward and uncouth expressions of countenance, as pouting, stretching, or twisting the lips; do not bite, smack, nor lick the lips; in enunciating emphatic words or sentences, do not pull down the corners of the mouth and expose the teeth as in grinning; the mouth should be used much more than the lips in forcible speaking.

In all proper gesticulation the movements of the body correspond with and express, in the language of signs, the thoughts and feelings of the speaker. This is done normally by young children, and by all persons who have not been perverted by miseducation. The tendency of the teachings of most of our schools is to exaggeration, by which the student acquires an artificial and affected mannerism. It is propriety, not quantity, of gesture that should be studied. The person who forgets himself in his subject seldom errs in gesticulation, while the person who puts himself before his subject always does. The question for the speaker, who would become proficient in gesture, to ask himself, is not, "What do the hearers think of *me?*" but "How do I present the subject?" If the speaker successfully communicates his thoughts and feelings to others, he will most certainly do himself justice in manner.

In standing, rest alternately on each foot, and principally on the heel, changing position frequently. Keep the feet always flat on the floor, avoiding all tendency to rest on the toes or on one edge of either foot. In walking the stage, turn by placing one foot behind the other, thus at all times inclining to face the audience; never make the awkward blunder of turning one foot around the other in front, thus bringing the back to the audience. The grace of oratorical action consists in the freedom and simplicity of those gestures which illustrate the subject.

On this subject the reader who aims at excellence will do well to read Pittenger's "Oratory, Sacred and Secular," which gives a history of some of the leading orators, preachers, and lecturers of the present day, and of the preceding century.

CHAPTER VI.

EXERCISES ON THE ELEMENTARY SOUNDS.

Those who would excel as speakers, readers, or singers, **should be able to** enunciate, distinctly and rapidly, all of the primary **or** elementary sounds which are represented by written language. The twenty-six letters of the English alphabet represent forty-four distinct sounds, as explained in the following table:

ANALYSIS OF THE ELEMENTARY SOUNDS.

There are forty-four **sounds** of the English **language,** represented by the twenty-six **letters** of the **alphabet and** their combinations, **as** in the following table:

1. a, long, as in ale, pale, national, plaintiff, amen.
2. a, grave, or Italian, as in ah, far, papa, mamma.
3. a, broad, or German, as in all, draw, daughter, fraught.
4. a, **short,** as in **at,** hat, attack, malefactor.
5. b, name sound, as in be, bite, bright, tub, hubbub
6. c, sound of s, as in cent, city, cornice, precipice.
7. c, **sound** of k, as in cap, come, occult, ecliptic.
8. c, sound of z, as in suffice, discern, sacrifice.
9. c, sound of sh, as in ocean, Phocion, Cappadocia.
10. d, name sound, as in ride, did, daddy, double-headed.
11. d, sound of t; as in faced, watched, dipped, **escaped.**
12. e, long, as in eel, peel, creed, reveal, precede.
13. e, short, as in ell, expel, ever-extended.
14. f, **name** sound, as in if, rife, fife, faithful, tariff.
15. f, **sound** of **v,** as in of, hereof, whereof, thereof.

16. g, soft or name sound, as in gem, ginseng, logical.
17. g, hard, as in go, give, gig, Brobdignag.
18. g, sound of gh, as in rouge, protege, mirage.
19. h, name sound, as in hale, high, Hannah.
20. i, long, as in isle, lilac, oblige, iodine.
21. i, short, as in in, pin, king, distinctive.
22. l, name sound, as in lo, lily, dalliance, lullaby.
23. m, name sound, as in map, mummy, amalgamate.
24. n, name sound, as in nine, ninny, nobleman, manikin.
25. n, sound of ng, as in bank, ingot, congress, angular.
26. o, long, as in old, osier, trophy, sofa, atrocious.
27. o, close, as in ooze, douceur, accoutre, troubadour.
28. o, short, as in on, combat, obelisk, holyday.
29. p, name sound, as in pill, pippin, panter, platter.
30. r, smooth, as in war, afar, tartar, murderer.
31. r, trilled, as in rough, railroad, recreation.
32. u, long, as in mute, astute, educate, judicature.
33. u, short, as in up, mum, ultra, numbskull.
34. u, full, as in pull, cruel, Prussian, Brutus.
35. w, name sound, as in woo, bewail, wigwam, wormwood.
36. x, name sound, as in axe, coxcomb, luxury, example.
37. x, sound of gz, as in exist, exhibit, exuberant.
38. y, name sound, as in ye, yoke, yewyaw, yesterday.
39. ch, name sound, as in charm, church, chickering, Chimborazo.
40. th, aspirate, as in thin, think, thankless, prothonotary.
41. th, vocal, as in than, that, beneath, withhold, wherewithal.
42. wh, name sound, as in what, wherefore, whirligig, whimpering.
43. oi or oy, diphthongs, or digraphs, as oil, boy, recoil, employ.
44. ou or ow, diphthongs, or digraphs, as in our, bow, gouty, trowel.

The student should master all of these sounds, and practice on them until he can repeat them with facility backward or forward; after which he may, with advantage, exercise on the different sounds or groups of sounds, with the view of developing the power of particular portions of the vocal and respiratory apparatus.

ANALYSIS OF THE SOUNDS OF LETTERS.

In order to ascertain the exact sound represented by any letter, character, or combination of letters, the student has only to analyze a word in which it occurs. The process is a very simple one, yet many teachers have never learned it.

Ask the scholar in the primary school, "How many sounds has b?" and he may answer promptly, "B has but one sound, as in bite." Very well; then ask him, "What *is* that sound of b, as in bite?" and he may not be able to tell you.

To ascertain what the sound of b is, and to be able to make it, pure and simple, he has only to analyze, vocally, any word or syllable containing the letter. It is more convenient for new beginners to take a word commencing with the letter, "as in bite." Let him spell and pronounce all the letters in the usual manner—b-i-t-e, bite. Then spell and pronounce all except the last letter, e—b-i-t, bit. The i being long, as in isle, the pronunciation of bite is precisely the same without the terminal e as with it; hence the scholar discovers that e is silent in that word. Next let him spell and pronounce the word, omitting the last two letters, t and e—b-i, bi. He now learns that i has its long sound in that word; if it were short it would be sounded like i in hit. Lastly let him sound the word omitting the last three letters. He will then enunciate the one sound of b, as in bite; and a little attention to the vocal organs will show him precisely how the sound of b is made.

The process of analysis is now completed; and by observing the position and action of the lips, he learns why the letter b belongs to the category of *labial* or lip sounds, its pronunciation, as well as that of m and p, requiring a closure of the lips.

By the application of this key the student can readily ascertain the sound of any letter or character.

Another similar and still more simple method is, to

select a word beginning with the letter or character the sound of which is to be ascertained; commence the pronunciation of the word, but stop the effort instantly with the first sound which the ear recognizes; this will be the pure sound by itself, whether vocal or aspirate.

Thus, if the student begin to pronounce the names, Cicero and Cato, and the words, this and thin, and interrupts the effort with the first appreciable noise, he will learn that c in Cicero has the hissing sound of s, and c in Cato the hard sound of k; while th in the word this, has a compound vocal sound, and th in thin, a compound aspirate or breath sound.

EXERCISES ON THE VOWEL SOUNDS.

There are sixteen vowel sounds in our language, including the diphthongs; they are found in the order of our alphabet in the following words: *ale, ah, all, at, eel, ell, isle, ill, old, ooze, on, use, up, full, oil, how.* The enunciation of these vowel sounds, distinct from that of the consonant sounds, in reading, speaking, and singing, is one of the best exercises for acquiring flexibility of the articulating muscles, and elasticity of the vocal cords; also for bringing into vigorous co-operative action those respiratory muscles which are most immediately concerned in the production of the lower tones of voice. They should be pronounced forward and backward until they can be repeated several times with a single respiration, thus:

ale, ah, all, at, eel, ell, isle, ill, old, ooze, on, use, up-full oil, how, a, a, a, a, e, e, i, i, o, o, o, u, u, w, oi, ow.

Reversely,

ow, oi, u, u, u, o, o, o, i, i, e, e, a, a, a, a, how, oil, full, up, use, on, ooze, old, ill, isle, ell, eel, at, all, ah, ale.

This exercise may be advantageously varied by em-

EXERCISES ON THE ELEMENTARY SOUNDS. 51

ploying only the short vowel sounds in the same manner.
at, ell, ill, on, up—up, on, ill, ell, at, a, e, i, o, u—u, o, i,
e, a.

Reading by the vowel sounds alone, is an exceedingly useful exercise for the articulating muscles, and may serve to " vary the entertainment." No better example for practice can be found than Hamlet's advice to the players.

Speak the speech, I pray you, as *I* pronounced it to you; *trippingly*
 e e o ,i a u, a i o ou i o u; i i i
on the tongue. But if you *mouth* it, as *many* of our players do, I had
o e u. u i u ou i, a a i o ou a e o, i a
as lief the town-*crier* had spoke my lines. And do not saw the *air* too
a e e ou i e a o i i . a o o a e a o
much with your *hand;* but use all *gently;* for in the very *torrent,*
u i u a ; u u a e i; o i e e i o e,
tempest, and, as I may say, WHIRLWIND of your passion, you must
e e, a , a i a a, i i o u a u, u u
acquire and beget a *temperance* that may give it *smoothness.* Oh! it
a i a e e a e a a i i o e . o i
offends me to the *soul* to hear a *robustious periwig-*pated fellow, tear a
o e e o e o o e a o u i u e i i a e o o, a a
passion to *tatters,* to very *rags,* to split the ears of the *groundlings.*
a u o a e , o o i a , o i e e o e o ·i .

EXERCISES ON THE CONSONANT SOUNDS.

There are seventeen *vocal* and eleven *aspirate* sounds in the English language; consonants are also distinguished into *simple,* of which there are thirteen, and *compound,* of which there are fifteen.

CONSONANTS. { *Vocal.*—b, as in bite; c, as in discern; d, as in dome; f, as in thereof; g, as in gem; g, as in go; g, as in menagerie; l, as in line; m, as in mamma; n, as in not; n, as in clank; r, as in jar; r, as in bright; w, as in wist; x, as in excite; y, as in youth; th, as in thee.

CONSONANTS.
{
Aspirate.—c, as in cent; c, as in cap; c, as in gracious; d, as in embraced; f, as in fit; h, as in hand; p, as in pop; x, as in extant; ch, as in chance; th, as in thin; wh, as in whine.

Simple.—b, as in bib; c, as in circle; c, as in Connecticut; d, as in day; d, as in tripp'd; f, as in foe; g, as in give; h, as in hope; l, as in live; m, as in man; n, as in ten; p, as in poppy; r, as in more.

Compound.—c, sound of z, as in suffice; c, sound of sh, as in judicial; f, sound of v, as in hereof; g, soft, as in ginger; g, sound of zh, as in tongue; n, sound of ng, as in Frank; r, rough or trilled, as in crash; w, name sound, as in wool; x, sound of ks, as in excel; x, sound of gz, as in example; y, name sound, as in yarn; ch, sound of tch, as in much; th, soft, or aspirate, as in theme; th, vocal, as in thou; wh, name sound, as in when.
}

Every consonant sound should be distinctly recognized and enunciated, until the whole list of twenty-eight can be repeated forward and backward with a single respiration. Exercises on the consonant sounds are calculated to promote rapidity and accuracy in the action of the tongue, lips, and mouth.

The following words represent the consonant sounds in the order heretofore mentioned: bob, cent, come, suffice, ocean, ride, dipped, rife, of, gem, go, mirage, hale, lo, man, nine, bank, pin, war, rough, wo, axe, exist, yoke, charm, thin, than, what.

By analyzing these words in the manner already explained, the sound represented by each letter or combination of letters will be readily ascertained.

EXERCISES IN EMPHASIS.

Stress.—The *first* three, and the *last* two verses, or volumes; not the *three* first and the *two* last; there can be only *one first* thing.

Quantity.—Roll on, thou dark and deep blúe ocean— roll! Ten thousand fleets sweep over thee in vain. Hail! —universal Lord.

Expulsive Stress.—Aim at nothing higher until you can read and speak deliberately, clearly, distinctly, and with the appropriate emphasis.

Stress and Higher Pitch.—O man, tyrannic lord! how long—how long, shall prostrate nature groan beneath your rage!

Prolongation and Monotone.—I appeal to you—O ye hills and groves of Alba, and your demolished altars! I call you to witness!—and thou—O holy Jupiter!

Rhetorical Pause.—Will all great Neptune's ocean wash this blood—clean—from my hands? No, these, my hands, will rather the multitudinous sea incarnadine, making the green—one red.

Change of the Seat of Accent.—Temperance and virtue raise men above themselves to angels; intemperance and vice sink them below themselves to the level of brutes.

SHOUTING.

Charge! Chester! charge! on Stanley, on;
Liberty, freedom—tyranny is dead;
Run hence; proclaim it in the streets—
The combat deepens! ON, *ye brave!*

EXAMPLES OF INTONATIONS.

Rising.—Are you desirous of becoming a good reader, speaker, and singer? Then learn and practice the principles herein taught and demonstrated.

Falling.—A mind properly disciplined to submit to a small present evil, to obtain a greater distant good, will often reap victory from defeat and honor from repulse.

Rising and Falling.—To whom the goblin, full of wrath, replied: Art thou traitor angel? Art thou he who first broke peace in heaven, and faith till then unbroken? Back to the punishment—false fugitive!

The man who is in the daily use of ardent spirits, if he does not become a drunkard, is in danger of losing his health and character.

EXAMPLES OF WAVES OR CIRCUMFLEXES.

Rising.—The love of approbation—produces excellent effects on men of sense; a strong desire for praise in weak minds conduces to little else than vanity.

Falling.—It is not prudent to trust your secrets to a man who can not keep his own. If you had made that affirmation, I might perhaps have believed it.

Combination.—Mere hirelings and time-servers—are always opposed to improvements and originality: so are tyrants—to liberty and republicanism.

CADENCE.

Ye nymphs of Solyma, begin the song;
To heavenly themes sublimer strains belong.

Such honors Ilion to her lover paid,
And peaceful slept the mighty Hector's shade.

EXAMPLES OF DYNAMICS.

Loud.—With mighty crash the noise astounds; amid Carnarvon's mountains rages loud, the repercussive roar; and Thule bellows through her utmost isles.

Rough.—The tempest growls; the unconquerable lightning struggles through, ragged and fierce, and— raging, strikes the aggravating rocks.

SOFT.

Soft roll your incense, herbs, and fruits, and flowers
Ye softer floods, that lead the humid maze
Along the vale. Breathe your still song
Into the reaper's heart.

SMOOTH.

Perfumes as of Eden flowed sweetly along,
And a voice as of angels enchantingly sung.

And the smooth stream in smoother numbers flowed.

Harsh.—On a sudden, open fly with impetuous recoil and jarring sound the infernal doors, and on their groaning hinges grate harsh thunder.

Forcible.—Now storming fury rose, and clamor, such as heard in heaven, till now, was never; arms on armor clashing, brayed horrible discord.

Harmonious.—As earth asleep, unconscious lies; effuse your mildest beams, ye constellations, while your angels strike, amid the spangled sky, the silver lyre.

Strong.—Him the Almighty power hurled headlong, flaming from the ethereal skies, with hideous ruin and combustion down to bottomless perdition.

CHAPTER VII.

SELECTIONS FOR PRACTICE.

THE student who aims at excellence in speaking or writing should carefully study, and become familiar with, the spirit of the masters of elocution and composition. He will profit more in studying well, practicing thoroughly, on a single production from one of their pens, than by memorizing and declaiming a hundred indifferent compositions by second-rate authors. Booth, Jefferson, Salvini, and Cushman, by mastering the characters of Hamlet, Rip Van Winkle, Othello, and Meg Merrilies, can have a profitable field of action for a life-time in playing those characters alone. One thing well done, in elocution as in other vocations, prepares the way for doing other things well, and leads the way to honor and prosperity.

In the following selections the **masters of** language and of oratory are represented, and **their productions** may not be excelled for ages. The selections are arranged with the view to public declamation as well as private **exercise.**

TO RANGE.

Strike home, strong-hearted man! down to the root
Of old oppression sink the Saxon steel.
Thy work is to hew down. In God's name, then,
Put nerve into thy task. Let other **men**

Plant, as they may, that better tree, whose fruit,
The wounded bosom of the church shall heal,
Be though the image-breaker. Let thy blows
Fall heavily as the Suabian's iron hand,
On crown or crosier, which shall interpose
Between thee and the weal of Father-land.
Leave creeds to closet idlers. First of all,
Shake thou all German dream-land with the fall
Of that accursed tree, whose evil trunk
Was spared of old by Erfart's stalwart monk.
Fight not with ghosts and shadows. Let us hear
The snap of chain-links. Let our gladdened ear
Catch the pale prisoner's welcome, as the light
Follows thy axe-stroke, through his cell of night.
Be faithful to both worlds; nor think to feed
Earth's starving millions with the husks of creed.
Servant of Him whose mission high and holy
Was to the wronged, the sorrowing, and the lowly,
Thrust not His Eden promise from our sphere,
Distant and dim beyond the blue sky's span;
Like him of Patmos, see it, now and here,—
The New Jerusalem comes to man!
Be warned by Luther's error. Nor like him,
When the routed Tuton dashes from his limb
The rusted chain of ages, help to bind
His hands, for whom thou claim'st the freedom of the mind!

GLORY.

1. The crumbling tombstone and the gorgeous mausole'um, the sculptured marble, and the venerable cathedral, all bear witness to the instinctive desire within us to be remembered by coming generations. But how short-lived is the immortality which the works of our hands can confer! The noblest monuments of art that the world has ever seen are covered with the soil of twenty centuries. The works of the age of Pericles lie

at the foot of the Acropolis in indiscriminate ruin. The plow-share turns up the marble which the hand of Phidias had chiseled into beauty, and the Mussulman has folded his flock beneath the falling columns of the temple of Minerva.

2. But even the works of our hands too frequently survive the memory of those who have created them. And were it otherwise, could we thus carry down to distant ages the recollection of our existence, it were surely childish to waste the energies of an immortal spirit in the effort to make it known to other times, that a being whose name was written with certain letters of the alphabet, once lived, and flourished, and died. Neither sculptured marble, nor stately column, can reveal to other ages the lineaments of the spirit; and these alone can embalm our memory in the hearts of a grateful prosperity.

3. As the stranger stands beneath the dome of St. Paul's, or treads, with religious awe, the silent aisles of Westminster Abbey, the sentiment, which is breathed from every object around him, is, the utter emptiness of sublunary glory. The fine arts, obedient to private affection or public gratitude, have here embodied, in every form, the finest conceptions of which their age was capable. Each one of these monuments has been watered by the tears of the widow, the orphan, or the patriot.

4. But generations have passed away, and mourners and mourned have sunk together into forgetfulness. The agèd crone, or the smooth-tongued beadle, as now he hurries you through ailes and chapel, utters, with measured cadence and unmeaning tone, for the thousandth time, the name and lineage of the once honored dead; and then gladly dismisses you, to repeat again his well-conned lesson to another group of idle passers-by.

5. Such, in its most august form, is all the immortality that matter can confer. It is by what we ourselves have done, and not by what others have done for us, that we shall be remembered by after ages. It is by thought that has aroused my intellect from its slumbers, which has "given lustre to virtue, and dignity to truth," or by those examples which have inflamed my soul with the love of goodness, and not by means of sculptured marble, that I hold **communion with** Shakespeare and Milton, with Johnson **and Burke, with** Howard and Wilberforce.
DR. WAYLAND.

CATO'S SOLILOQUY.

1 It must **be so—Plato,** thou reasonest well!
Else, whence this pleasing hope, this fond desire,
This longing after immortality?
Or whence this secret dread, and inward horror,
Of falling into naught? **Why** shrinks the soul
Back **on herself, and startles at** destruction?
'Tis the divinity **that stirs within us;**
'Tis Heaven itself that points out a **hereafter,**
And intimates eternity to man.

2 Eternity!—though pleasing, dreadful thought!
Through what variety of untried being,
Through what new scenes and changes must we pass!
The wide, **the** unbounded prospect lies before me;
But shadows, **clouds,** and darkness rest upon it.
Here will **I hold.** If there's a Power above us,—
And that **there is, all Nature cries aloud**
Through all her **works,—He must delight in virtue;**
And that which **He delights in** must be happy.
But when? or where? **This** world was made for Cæsar.
I'm weary of conjectures,—this must **end them.**
[Laying his hand on his sword.

3 Thus am I doubly armed. My **death** and life,
My bane and antidote, are both before me.

This in a moment brings me to my end;
But this informs me I shall never die.
The soul, secure in her existence, smiles
At the drawn dagger, and defies its point.
The stars shall fade away, the sun himself
Grow dim with age, and Nature sink in years;
But thou shalt floŭrish in immortal youth,
Unhurt amid the war of elements,
The wreck of matter, and the crush of worlds.

ADDISON.

OUR HONORED DEAD.

1. How bright are the honors which await those who with sacred fortitude and pātriŏt′ic patience have endured all things that they might save their native land from dĭvĭsion and from the power of corruption! "The honored dead! They that die for a good cause are redeemed from death. Their names are gáthered and garnered. Their memory is precious. Each place grows proud for them who were born there.

2. There is to be, ere lŏng, in ĕvĕry village and in every neighborhood, a glowing pride in its martyred heroes. Tablets shall preserve their names. Pious love shall renew their inscriptions as time and the unfeeling elements decay them. And the nătional festivals shall give multitudes of precious names to the ŏrator's lips. Children shall grow up under mōre sacred inspirations, whose elder brothers, dying nobly for their country, left a name that honored and inspired all who bore it. Orphan children shall find thousands of fathers and mothers to love and help those whom dying heroes left as a legacy to the gratitude of the public.

3. Oh, tell me not that they are dead—that generous hōst, that airy army of invisible heroes! They hover as a cloud of witnèssès above this nation. Are they dead

that yĕt speak louder than we can speak, and a mōre universal language? Are they dead that yet act? Are they dead that yet move upon society, and inspire the people with nobler motives and more heroic pātriötĭsm?

4. Ye that mōurn, let gladnèss mingle with your tears. He *was* your son; but now he *is* the nation's. He made your household bright: now his example inspires a thousand households. Dear to his brothers and sisters, he is now brother to ĕvèry generous youth in the land. Before, he was nărrowed, appropriated, shut up to you. Now he is augmented, set free, and given to all. He has died from the fămily, that he might live to the nation. Not one name shall be forgotten or neglected; and it shall by-and-by be confessed, as of an āncient hērō, that he did more for his country by his death than by his whōle life.

5. Nĕither are they less honored who shall bear through life the marks of wounds and sufferings. Neither ĕp′aŭlĕtte nor badge is so honorable as wounds received in a good cause. Many a man shall envy him who henceförth limps. So strānge is the transforming power of pātriŏtic ardor, that men shall almost covet disfigurement. Crowds will give way to hobbling cripples, and uncover in the presence of feeblenèss and helplessness. And buoyant children shall pause in their noisy games, and with loving reverence honor them whose hands can work no more, and whose feet are no longer able to march except upon that journey which brings good men to honor and immortality.

6. O mother of lŏst children! set not in darknèss nor sorrow whom a nation honors. O mōurners of the early dead! they shall live again, and live forever. Your sórrows are our gladness. The nation lives, because you

gave it men that loved it better than their own lives. And when a few mōre days shall have cleared the pĕrils from around the nation's brow, and she shall sit in unsullied garments of liberty, with justice upon her fŏre*h*ĕad, love in her eyes, and truth upon her lips, she shall not forget those whose blood gave vital cŭrrents to her heart, and whose life, given to her, shall live with her life till time shall be no more.

7. Every mountain and hill shall have its treasured name, every river shall keep some solemn title, every valley and every lake shall cherish its honored register; and till the mountains are wōrn out, and the rivers forget to flow, till the clouds are weary of replenishing springs, and the springs forget to gush, and the rills to ŝing, shall their names be kept fresh with reverent honors, which are inscribed upon the book of National Remembrance!

H. W. Beecher.

DARKNESS.

1. I had a dream, which was not all a dream.
The bright sun was extinguished, and the stars
Did wander, darkling, in the eternal space,
Raylĕss and pathless, and the icy earth
Swung blind and blackening in the moonless air.
Morn came, and went—and came, and brought no day,
And men forgot their passions, in the dread
Of this their desolation; and all hearts
Were chilled into a selfish prayer for light.
And they did live by watch-fires; and the thrones,
The palaces of crownèd kings, the huts,
The habitations of all things which dwell,
Were burnt for bĕacons: cities were consumed,
And men were gathered round their blazing homes,
To look once mōre into each other's face.
Happy were those who dwelt within the eye
Of the volcanoes and their mountain torch.

SELECTIONS FOR PRACTICE.

2. A fearful hope was all the world contained:
Forests were set on fire; but hour by hour,
They fell and faded; and the crackling trunks
Extinguished with a crash—and all was black.
The brows of men, by their despairing light,
Wore an unearthly aspect, as, by fits,
The flashes fell upon them. Some lay down,
And hid their eyes, and wept; and some did rest
Their chins upon their clenchèd hands, and smiled;
And others hurried to and fro, and fed
Their funeral piles with fuel, and looked up,
With mad disquietude, on the dull sky,
The pall of a past world; and then again
With curses, cast them down upon the dust,
And gnashed their teeth, and howled. The wild birds
    shrieked,
And, terrified, did flutter on the ground,
And flap their useless wings: the wildest brutes
Came tame and tremulous; and vipers crawled
And twined themselves among the multitude,
Hissing, but stingless—they were slain for food.

3. And War, which for a moment was no more,
Did glut himself again:—a meal was bought
With blood, and each sat sullenly apart,
Gorging himself in gloom; no love was left;
All earth was but one thought—and that was death,
Immediate and inglorious; and the pang
Of famine fed upon all entrails. Men
Died; and their bones were tombless as their flesh.
The meager by the meager were devoured.
Even dogs assailed their masters,—all save one,
And he was faithful to a corpse, and kept
The birds, and beasts, and famished men at bay,
Till hunger clung them or the drooping dead
Lured their lank jaws: himself sought out no food,
But with a piteous and perpetual moan,
And a quick, desolate cry, licking the hand
Which answered not with a caress—he died.

4. The crowd was famished by degrees. But two
   Of an enormous city did survive,
   And they were enemies. They met beside
   The dying embers of an altar-place,
   Where had been heaped a mass of holy things
   For an unholy usage. They raked up,
   And, shivering, scraped with their cold skeleton hands,
   The feeble ashes, and their feeble breath
   Blew for a little life, and made a flame,
   Which was a mŏckery. Then they lifted
   Their eyes as it grew lighter, and beheld
   Each other's aspects—saw, and shrieked, and died;
   Even of their mutual hideousness they died,
   Unknowing who he was upon whose brow
   Famine had written Fiend.

5.   The world was void:
   The populous and the powerful was a lump,
   Seasonless, herbless, treeless, manless, lifeless;
   A lump of death, a chaos of hard clay.
   The rivers, lakes, and ocean all stood still,
   And nothing stirred within their silent depths.
   Ships, sailorless, lay rotting on the sea,
   And their masts fell down piecemeal: as they dropped
   They slept on the abyss, without a surge,—
   The waves were dead ; the tides were in their grave;
   The moon, their mistress, had expired before;
   The winds were withered in the stagnant air,
   And the clouds perished : Darkness had no need
   Of aid from them—she was the universe.
                                           LORD BYRON.

A CURTAIN LECTURE OF MRS. CAUDLE.

Bah! that's the third umbrella gone since Christmas. What were you to do? Why, let him go home in the rain, to be sure. I'm very certain there was nothing about him that could spoil.—Take cold, indeed! He doesn't look like one of the sort to take cold. Besides he'd have better **taken cold** than taken our umbrella.—Do you hear the rain, Mr. Caudle? I say, do you hear the rain? And, as I'm alive, if it isn't St. Swithin's day! Do you hear it against the windows? Nonsense! you don't impose upon me; you can't be asleep **with such a shower as that!** Do you hear it, I say? Oh! **you DO hear it!** Well, that's a pretty flood, I think, to last for six weeks; and no stirring all the time out of the house.

2. Pooh! don't think **me a fool**, Mr. Caudle; don't insult me; *he* return the umbrella! Anybody would think you were born yĕsterday. As if anybody ever did return an umbrella! There: do you hear it? **Worse and worse.** Cats and dŏgs, **and for six** weeks: always **six weeks;** and no umbrella!—I should **like to know** how the children are to go **to** school to-mŏrrōw! They **shan't** go through such weather; I am determined. No; they shall stop at home and never learn any thing (the blessèd creatures!) sooner than **go and** gĕt **wet!** And when they **grow up, I** wonder who they'll have to **thank** for knowing **nothing**: who, indeed, but their **father.** People who can't feel for their own **children ought never** to be fathers.

3. But I know why you lent the umbrella: oh! yes, I know vĕry **well.** I was going out to tea at dear mother's to-mŏrrōw: **you knew that,** and you did it on purpose. Don't tell me; you hate to have me go there, and take every mean **advantage to hinder** me. But don't **you** think it, Mr. Caudle; no, sir: if it comes down in bucketfulls, I'll **go all the mōre.** No; and I wōn't have a cab! Where do you **think the money's to come from?** You've got nice high notions **at that club** of yours! A cab, indeed! Cŏst me sixteen pence, at least. Sixteen pence! two-and-eight pence; **for there's** back again. Cabs, indeed! I should like to know who's to pay for 'em; for

I'm sure you can't, if you go on as you do, throwing away your property, and beggaring your children, buying umbrellas!

4. Do you hear the rain, Mr. Caudle? I say, do you hear it? But I don't care—I'll go to mother's to-mŏrrōw—I will; and what's more, I'll walk every step of the way; and you know that will give me my death. Don't call me a foolish woman; it's you that's the foolish man. You know I can't wear clogs; and with no umbrella, the wet's sure to give me a cold: it always does; but what do you care for that? Nothing at all. I may be laid up, for what you care, as I dare say I shall; and a pretty doctor's bill there'll be. I hope there will. It will teach you to lend your umbrellas again. I shouldn't wonder if I caught my death: yĕs, and that's what you lent the umbrella for. Of cōurse!

5. Nice clothes I gĕt, too, traipsing through weather like this. My gown and bŏnnet will be spoiled quite. Needn't I wear 'em, then? Indeed, Mr. Caudle, I *shall* wear 'em. No, sir; I'm not going out a dowdy, to please you, or anybody else. Gracious knows! it isn't ŏften that I step over the threshold:—indeed, I might as well be a slave at once: better, I should say; but when I do go out, Mr. Caudle, I choose to go as a lady.

6. Oh! that rain—if it isn't enough to break in the windōws. Ugh! I look forward with dread for to-mŏrrōw! How am I to go to mother's, I'm sure I can't tell; but if I die, I'll do it.—No, sir; I wŏn't bŏrrōw an umbrella: no; and you shan't *buy* one. Mr. Caudle, if you bring home another umbrella, I'll throw it into the street. Ha! And it was ōnly last week I had a new nozzle put to that umbrella. I'm sure if I'd have known as much as I do now, it might have gōne without one. Paying for new

nozzles for other people to laugh at you! Oh! it's all vĕry well for you; you can go to sleep. You've no thought of your poor patient wife, and your own dear children; you think of nothing but lending umbrellas! Men, indeed!—call themselves lords of the creätion! pretty lords, when they can't even take care of an umbrella!

7. I know that walk to-mŏrrōw will be the death of me. But that's what you want: then you may go to your club, and do as you like; and then nicely my poor dear children will be used; but then, sir, then you'll be happy. Oh! don't tell me! I know you will: else you'd never have lent the umbrella!—You have to go on Thursday about that summons; and, of cōurse, you can't go. No, indeed: you *don't* go without the umbrella. You may lose the debt, for what I care—it wōn't be so much as spoiling your clothes—better lose it: people deserve to lose debts who lend umbrellas!

8. And I should like to know how I'm to go to mother's without the umbrella. Oh! don't tell me that I said I *would* go; that's nothing to do with it,—nothing at all. She'll think I'm neglecting her; and the little money we're to have, we shan't have at all;—because we've no umbrella.—The children, too! (dear things!) they'll be sopping wet: for they shan't stay at home; they shan't lose their learning; it's all their father will leave them, I'm sure! But they *shall* go to school. Dōn't tell me they shouldn't (you are so aggravating, Caudle, you'd spoil the temper of an āngel!); they *shall* go to school: mark that! and if they get their deaths of cold, it's not my fault; I DIDN'T LEND THE UMBRELLA.   JERROLD.

IMMORTALITY.

"Man, thou shalt never die!" Celestial voices
Hymn it unto our souls: according harps,

By ängel fingers touched, when the mild **stars**
Of morning **sang** together, sound förth still
The sŏng **of our great** immortality!
Thick-clustering orbs on this our fair domain,
The tall, dark mountains, and the deep-tōned seas,
Join in this solemn, universal song.
O lis*t*en, ye our spirits! drink it in
From all the air! 'Tis in the gentle moonlight;
'Tis flōating 'mid day's setting glōries; night,
Wrapped in her sable robe, with silent step
Comes to our bed, and breathes it in our ears.
Night and the dawn, bright day and thoughtful eve,
All time, all bounds, the limitlèss expanse,
As one vast mўstic instrument, are touched
By an unseen, living hand, and conscious chords
Quiver with joy in this great jubilee:
The dying hear it; and, as sounds of earth
Grow dull and distant, wake their passing souls
To mingle in this heavenly harmony."

R. H. DANA.

ADVANTAGES OF ADVERSITY.

1. From the dark pōrtals of the stär-chämber, and in the stern text of the acts of uniformity, the Pilgrims received a commission, mōre efficient than any that ever bōre the royal seal. Their banishment to Holland was fortunate; the decline of their little company in the strānge land was fortunate; the difficulties which they experienced in gĕtting the royal consent to banish themselves to this wĭldernèss were fortunate; all the tears and heart-breakings of that memorable parting at Delfthaven had the happiëst influence on the rising destinies of New England. All this purified the ranks of the settlers. These rough touches of fortune brushed ŏff the light, uncertain, selfish spirits. They made it a grave, solemn, self-denying expedition, and required of those who en-

gaged in it to be so too. They cast a broad shadōw of thought and seriousness over the cause; and, if this sometimes deepened into melancholy and bitterness, can we find no apology for such a human weakness?

2. It is sad, indeed, to reflect on the disasters which the little band of Pilgrims encountered; sad to see a pōrtion of them, the prey of unrelenting cupidity, treacherously embarked in an unsound, unseaworthy ship, which they are soon oblīged to abandon, and crowd themselves into one vessel; one hundred persons, besides the ship's company, in a vessel of one hundred and sixty tons. One is touched at the stōry of the lŏng, cold, and weary autumnal passage; of the landing on the inhŏspital rocks at this dismal season; where they are deserted, before long, by the ship which had brought them, and which seemed their ōnly hold upon the world of fellōw-men, a prey to the elements and to want, and fearfully ignorant of the numbers, the power, and the temper of the savage tribes that filled the unexplōred continent upon whose verge they had ventured.

3. But all this wrought togĕther for good. These trials of wandering and exile, of the ocean, the winter, the wilderness, and the savage foe, were the final assurances of success. It was these that put far away from our fathers' cause all patrician sŏftness, all hereditary claims to preëminence. No effeminate nobility crowded into the dark and austere ranks of the Pilgrims. No Carr nor Villers would lead on the ill-provided band of despised Puritans. No well-endowed clergy were on the alert to quit their cathedrals, and set up a pŏmpous hierarchy in the frozen wilderness. No craving governors were anxious to be sent over to our cheerless El Dorādos of ice and snow.

4. No; they could not say they had encouraged, pătronized, or helped the Pilgrims: their own cares, their own labors, their own councils, their own blood, contrived all, achieved all, bōre all, sealed all. They could not afterwards fairly pretend to reap where they had not strewn; and, as our fathers reared this broad and solid fabric with pains and watchfulness, unaided, barely tolerated, it did not fall when the favor, which had always been withholden, was changed into wrath; when the arm, which had never supported, was raised to destroy.

5. Methinks I see it now, that one solitary, adventurous vessel, the Mayflower of a forlorn hope, freighted with the prospects of a future State, and bound ăcrōss the unknown sea. I behold it pursuing, with a thousand misgivings, the uncertain, the tedious voyage. Suns rise and set, and weeks and months pass, and winter surprises them on the deep, but brings them not the sight of the wished-for shōre.

6. I see them now scantily supplied with provisions; crowded almost to suffocation in their ill-stōred prison; delayed by calms, pursuing a circuitous route—and now driven in fury before the raging tĕmpèst, on the high and giddy waves. The awful voice of the storm howls through the rigging. The laboring masts seem straining from their base; the dismal sound of the pumps is heard; the ship leaps, as it were, madly, from billōw to billow; the ocean breaks and settles with ingulfing floods over the floating deck, and beats, with deadening, shivering weight, against the staggered vessel.

7. I see them, escaped from these perils, pursuing their all but desperate undertaking, and landed at last, after a five months' passage, on the ice-clad rocks of Plymouth—weak and weary from the voyage, poorly armed, scantily

provisioned, depending on the charity of their shipmaster for a draught of beer on bōard, drinking nothing but water on shōre—without shelter, without means—surrounded by hŏstĭle tribes.

8. Shut now the volume of history, and tell me, on any principle of human probability, what shall be the fate of this handful of adventurers. Tell me, man of military science, in how many months were they all swept ŏff by the thirty savage tribes, enumerated within the early limits of New England? Tell me, politician, how lŏng did this shadōw of a colony, on which your conventions and treaties had not smiled, languish on the distant cōast? Student of history, compare for me the baffled prŏjects, the deserted settlements, the abandoned adventures of other times, and find the parallel of this.

9. Was it the winter's storm, beating upon the houseless heads of women and children; was it hard labor and spare meals; was it disease; was it the tomahawk; was it the deep malady of a blighted hope, a ruined enterprise, and a broken heart, aching in its last moments at the recollection of the loved and left beyŏnd the sea—was it some, or all of these united, that hŭrried this forsaken company to their melancholy fate; and is it possible that nēither of these causes, that not all combined were able to blast this bud of hope? Is it possible, that, from a beginning so feeble, so frail, so worthy, not so much of admiration as of pity, there has gŏne fōrth a prŏgress so steady, a growth so wonderful, an expansion so ample, a reälity so important, a promise, yĕt to be fulfilled, so glōrious?

EDWARD EVERETT.

MORNING.

Sweet is the breath of Morn, her rising sweet,
  With charm of earliĕst birds; pleasant the sun,

When first on this delightful land he spreads
His orient beams, on herb, tree, fruit, and flower,
Glistening with dew; fragrant the fertile earth
After sŏft showers; and sweet the coming on
Of grateful evening mild: then silent Night,
With this her solemn bird, and this fair moon,
And these the gems of heaven, her starry train.
MILTON.

THE DILEMMA.—SCENE FROM PICKWICK.

Mr. Pickwick's apartments in Goswell street, although on a limited scale, were not ōnly of a vĕry neat and comfortable description, but peculiarly adapted for the residence of a man of his genius and observation. His sitting-room was the first floor front, his bed-room was the second floor front; and thus, whether he was sitting at his desk in the parlor, or standing befŏre the dressing-glass in his dormitory, he had an equal opportunity of contemplating human nature in all the numerous phases it exhibits, in that not mōre populous than popular thoroughfare.

2. His landlady, Mrs. Bardell—the relict and sole executrix of a deceased custom-house ŏfficer—was a comely (kŭm′ly) woman of bustling manners and agreeable appearance, with a natural genius for cooking, improved by study and lōng practice into an ĕx′quĭsĭte talent. There were no children, no servants, no fowls. The ōnly other inmates of the house were a large man and a small boy; the first a lodger, the second a production of Mrs. Bardell's. The large man was always at home precisely at ten o'clock at night, at which hour he regularly condensed himself into the limits of a dwarfish French bedstead in the back parlor; and the infantine spŏrts and gymnastic exercises of Master Bardell were exclusively

SELECTIONS FOR PRACTICE. 73

confined to the neighboring pavements and gutters. Clĕanliness and quiet reigned throughout the house; and in it Mr. Pickwick's will was law.

3. To any one acquainted with these points of the domestic economy of the establishment, and cŏn′versant with the admirable regulation of Mr. Pickwick's mind, his appearance and behavior, on the morning previous to that which had been fixed upon for the journey to Eatansvill, would have been mōst mysterious and unaccountable. He paced the room to and fro with hŭrried steps, popped his head out of the windōw at intervals of about three minutes each, constantly referred to his watch, and exhibited many other manifestations of impatience, vĕry unusual with him. It was evident that something of great importance was in contemplation; but what that something was, not even Mrs. Bardell herself had been enabled to discover.

4. "Mrs. Bardell," said Mr. Pickwick, at last, as that amiable female approached the termination of a prolonged dusting of the apartment. "Sir," said Mrs. Bardell. "Your little boy is a vĕry long time gone." "Why, it's a good long way to the Borough, sir," remonstrated Mrs. Bardell. "Ah," said Mr. Pickwick, "vĕry trúe; so it is." Mr. Pickwick relapsed into silence, and Mrs. Bardell resumed her dusting.

5. "Mrs. Bardell," said Mr. Pickwick, at the expiration of a few minutes. "Sir," said Mrs. Bardell again. "Do you think it's a much greater expense to keep two people, than to keep one?" "La, Mr. Pickwick," said Mrs. Bardell, coloring up to the vĕry border of her cap, as she fancied she observed a species of matrimonial twinkle in the eyes of her lodger; "La, Mr. Pickwick, what a question!" "Well, but *do* you?" inquired Mr. Pickwick. "That de-

4

pends," said Mrs. Bardell, approaching the duster very near to Mr. Pickwick's elbow, which was planted on the table; "that depends a good deal upon the person, you know, Mr. Pickwick ; and whether it's a saving and careful person, sir." " That's very true," said Mr. Pickwick ; " but the person I have in my eye (here he looked very hard at Mrs. Bardell) I think possesses these qualities; and has, moreover, a considerable knowledge of the world, and a great deal of sharpness, Mrs. Bardell; which may be of material use to me."

6. " La, Mr. Pickwick," said Mrs. Bardell ; the crimson rising to her cap-border again. " I do," said Mr. Pickwick growing energetic, as was his wont (wŭnt) in speaking of a subject which interested him. " I do, indeed ; and to tell you the truth, Mrs. Bardell, I have made up my mind." " Dear me, sir," exclaimed Mrs. Bardell. " You'll think it not very strange now," said the amiable Mr. Pickwick, with a good-humored glance at his companion, " that I never consulted you about this matter, and never mentioned it, till I sent your little boy out this morning—eh ?"

7. Mrs. Bardell could only reply by a look. She had lŏng worshipped Mr. Pickwick at a distance, but here she was, all at once, raised to a pinnacle to which her wildĕst and mōst extravagant hopes had never dared to aspire. Mr. Pickwick was going to propose—a deliberate plan, too—sent her little boy to the Borough, to get him out of the way—how thoughtful—how considerate !—" Well," said Mr. Pickwick, " what do you think ?" " Oh, Mr. Pickwick," said Mrs. Bardell, trembling with agitation, " you're vĕry kind, sir." " It will save you a great deal of trouble, wōn't it ?" said Mr. Pickwick. " Oh, I never thought anything of the trouble, sir," replied Mrs. Bar-

dell; "and of course, I should take more trouble to please you than ever; but it is so kind of you, Mr. Pickwick, to have so much consideration for my loneliness."

8. "Ah, to be sure," said Mr. Pickwick; "I never thought of that. When I am in town, you'll always have somebody to sit with you. To be sure, so you will." "I am sure I ought to be a very happy woman," said Mrs. Bardell. "And your little boy—" said Mr. Pickwick. "Bless his heart," interposed Mrs. Bardell, with a maternal sob. "He, too, will have a companion," resumed Mr. Pickwick, "a lively one, who'll teach him, I'll be bound, more tricks in a week, than he would ever learn in a year." And Mr. Pickwick smiled placidly.

9. "Oh you dear—" said Mrs. Bardell. Mr. Pickwick started. "Oh you kind, good, playful dear," said Mrs. Bardell; and without more ado, she rose from her chair, and flung her arms round Mr. Pickwick's neck, with a cataract of tears, and a chorus of sobs. "Bless my soul," cried the astonished Mr. Pickwick;—"Mrs. Bardell, my good woman—dear me, what a situation—pray consider. Mrs. Bardell, dŏn't—if anybody should come—" "Oh, let them come," exclaimed Mrs. Bardell, frantically; "I'll never leave you—dear, kind, good soul;" and, with these words, Mrs. Bardell clung the tighter.

10. "Mercy upon me," said Mr. Pickwick, struggling violently, "I hear somebody coming up the stairs. Dŏn't, don't, there's a good creature, don't." But entreaty and remonstrance were ălīke unavailing: for Mrs. Bardell had fainted in Mr. Pickwick's arms; and befŏre he could gain time to deposit her on a chair, Master Bardell entered the room, ushering in Mr. Tupman, Mr. Winkle, and Mr. Snodgrass. Mr. Pickwick was struck motionless and speechless. He stood with his lovely burden in

his arms, gazing vacantly on the countenances of his friends, without the slightest attempt at recognition or explanation. They, in their turn, stared at him; and Master Bardell, in his turn, stared at everybody.

11. The astonishment of the Pickwickians was so absorbing, and the perplexity of Mr. Pickwick was so extreme, that they might have remained in exactly the same relative situation until the suspended animation of the lady was restored, had it not been for a most beautiful and touching expression of filial affection on the part of her youthful son. Clad in a tight suit of corduroy, spangled with brass buttons of a very considerable size, he at first stood at the door astounded and uncertain; but by degrees, the impression that his mother must have suffered some personal damage, pervaded his partially developed mind, and considering Mr. Pickwick the aggressor, he set up an appalling and semi-earthly kind of howling, and butting forward, with his head, commenced assailing that immortal gentleman about the back and legs, with such blows and pinches as the strength of his arm and the vīölence of his excitement allowed.

12. "Take this little villain ăwāy," said the agonized Mr. Pickwick, "he's mad." "What *is* the matter?" said the three tongue-tied Pickwickians. "I don't know," replied Mr. Pickwick, pettishly. "Take away the boy—(here Mr. Winkle carried the in′teresting boy, screaming and struggling, to the farther end of the apartment). Now help me to lead this woman down stairs." "Oh, I'm better now," said Mrs. Bardell, faintly. "Let me lead you down stairs," said the ever gallant Mr. Tupman. "Thank you, sir—thank you;" exclaimed Mrs. Bardell, hysterically. And down stairs she was led accordingly, accompanied by her affectionate son.

13. "I can not conceive"—said Mr. Pickwick, when his friend returned—"I can not conceive what has been the matter with that woman. I had merely announced to her my intention of keeping a man-servant, when she fell into the extraordinary paroxysm in which you found her. Věry extraordinary thing." "Very," said his three friends. "Placed me in such an extremely awkward situation," continued Mr. Pickwick. "Very;" was the reply of his followers, as they coughed slightly, and looked dubiously at each other.

14. This behavior was not lost upon Mr. Pickwick. He remarked their incredulity. They evidently suspected him. "There is a man in the passage now," said Mr. Tupman. "It's the man that I spoke to you about," said Mr. Pickwick, "I sent for him to the Borough this morning. Have the goodness to call him up, Mr. Snodgrass." DICKENS.

DEITY.

1. A million torches, lighted by Thy hand,
Wander unwearied through the blue abyss—
They own Thy power, accomplish Thy command,
All gay with life, all eloquent with bliss.
What shall we call them? Piles of crystal light—
A glōrious company of golden streams—
Lamps of celestial ether burning bright—
Suns lighting systems with their joyous beams?
But Thou to these art as the noon to night.

2. Yĕs! as a drop of water in the sea,
All this magnificence in Thee is lost:—
What are ten thousand worlds compared to Thee?
And what am I then?—Heaven's unnumbered host,
Though multiplied by myriads, and arrayed
In all the glōry of sublimèst thought,
Is but an atom in the balance, weighed

Against Thy greatness—is a cipher brought
Against infinity! What am I then? Naught!

3. Naught! But the effluence of Thy light divine,
Pervading worlds, hath reached my bosom too;
Yĕs! In my spirit doth Thy spirit shine
As shines the sun-beam in a drop of dew.
Naught! but I live, and on hope's pinions fly
Eager toward Thy presence; for in Thee
I live, and breathe, and dwell; aspiring high.

4. Thou art!—dĭrĕcting, guiding all—Thou art!
Direct my understanding then to Thee;
Control my spirit, guide my wandering heart;
Though but an atom 'midst immensity,
Still I am something, fashioned by Thy hand!
I hold a middle rank 'twixt heaven and earth—
On the last verge of mortal being stand,
Close to the realms where angels have their birth,
Just on the boundaries of the spirit-land!

5. The chain of being is complete in me—
In me is matter's last gradation lost,
And the next step is spirit—Deity!
I can command the lightning, and am dust!
A monarch and a slave—a worm, a god!
Whence came I here, and how? so marvelously
Constructed and conceived? unknown! this clod
Lives surely through some higher energy;
For from itself alone it could not be!

6. Creator, yĕs! Thy wisdom and Thy word
Created me! Thou sōurce of life and good!
Thou spirit of my spirit, and my Lord!
Thy light, Thy love, in their bright plenitude
Filled me with an immortal soul, to spring
Over the abyss of death; and băde it wear
The garments of eternal day, and wing
Its heavenly flight beyŏnd this little sphere,
Even to its sōurce—to Thee—its Author there.

7. O thoughts ineffable! O visions blest!
Though worthless our conceptions all of Thee,
Yĕt shall Thy shadōwed image fill our breast,
And waft its homage to Thy Deity.
Gŏd! thus ălōne my lowly thoughts can sōar,
Thus seek Thy presence—Being wise and good!
'Midst Thy vast works admire, obey, ădōre;
And when the tongue is eloquent no mōre
The soul shall speak in tears of gratitude.
DERZHAVEN.

THE DEATH OF HAMILTON.

A short time since, and he, who is the occasion of our sŏrrōws, was the ornament of his country. He stood on an eminence, and glōry covered him. From that eminence he has fallen: suddenly, **forever fallen**. His intercōurse with the living world is now ended; and those who would hereafter find him, must seek him in the grave. There, cŏld and lifèless, is the heart which just now was the seat of friendship; there, dim and sightless, is the eye, whose rădiant and enlivening orb beamed with intelligence; and there, closed forever, are those lips, on whose persuasive accents we have so ŏft*e*n, and so lately hung with transpōrt!

2. From the **darkness** which rests upon his tomb there proceeds, methinks, a **light**, in which it is clearly seen, that those gaudy objects which men pursue are ōnly phantoms. In this light how dimly shines the splendor of victory—how humble appears the majesty of grandeur! The bubble, which seemed to have so much solidity, has burst; and we again see, that all belōw the sun is vanity!

3. True, the funeral eulogy has been pronounced, the sad and solemn procession has moved, the badge of mōurn-

ing has already been decreed, and presently the **sculptured marble** will lift up its front, proud to perpetuate the **name of** Hamilton, and rehearse to the passing traveler his virtues (just **tributes of** respect, and **to the** living useful); but to him, moldering in his **nărrōw and** humble habitation, what are they? How vain! how unavailing!

4. Approach, and behold, while I lift from his sepulchre its covering! Ye admirers of his greatnèss! ye emulous of his talents and his fame! approach and behold him now. How pale! how silent! No martial bands admire the adroitness of his movemènts; no fascinating thrŏng weep, and melt, and tremble at his eloquence! Amazing chănge! a shroud! a coffin! a nărrōw, subterraneous cabin!—this is all that now remains of Hamilton! And is this all that remains of Hamilton? During a life so transitory, what lasting monument, then, can our fondèst hopes erect!

5. My brethren, we stand on the borders of an awful gulf, which is swallōwing up all things human. And is there, amidst this universal wreck, nothing stable, nothing abiding, nothing immortal, on which poor, frail, dying man can fasten? Ask the hero, ask the statesman, whose wisdom you have been accustomed to revere, and he will tell you: He will tell you, did I say? He has already told you, from his death-bed; and his illumined spirit still whispers from the heavens, with well-known eloquence, the solemn admonition : "Mortals hastening to the tomb, and once the companions of my pilgrimage, take warning and avoid my errors; cultivate the virtues I have recommended; choose the Saviour I have chosen; live disinterestedly; live for immortality; and would you rescue any thing from final dissolution, lay it up in Gōd."

Nott.

THE STARS.

Roll on, ye stars; exult in youthful prime;
Mark with bright curves the printless steps of Time;
Near and more near your beamy cars approach,
And lessening orbs on lessening orbs encroach.
Flowers of the sky, ye, too, to age must yield,
Frail as your silken sisters of the field.

Star after star from heaven's high arch shall rush,
Suns sink on suns, and systems systems crush,
Headlŏng, extinct, to one dark centre fall,
And death, and night, and chaos mingle all;
Till ō'er the wreck, emerging from the storm,
Immortal Nature lifts her chāngeful form,
Mounts from her funeral pyre, on wings of flame,
And sōars and shines, another and the same.
Darwin.

PUBLIC VIRTUE.

1. I hope, that in all that relates to personal firmness, all that concerns a just appreciation of the insignificance of human life,—whatever may be attempted to threaten or alarm a soul not easily swayed by opposition, or awed or intimidated by mĕnace,—a stout heart ănd a steady eye, that can survey', unmoved and undaunted, any mere personal perils that assail this poor, transient, perishing frame, —I may, without disparagement, compare with other men.

2. But there is a sort of coŭrage, which, I frankly confess it, I do not possess,—a bŏldnèss to which I dare not aspire, a valor which I can not covet. I can not lay myself down in the way of the welfare and happiness of my country. That I can not, I have not the courage to do.

I can not interpose the power with which I may be invested—a power conferred, not for my personal benefit, nor for my aggrăn´dīzement, but for my country's good— to check her onward march to greatness and glōry. I have not courage enough. I am too cowardly for that.

3. I would not, I dare not, in the exercise of such a trust, lie down, and place my body ăcrŏss the path that leads my country to prosperity and happiness. This is a sort of coŭrage widely different from that which a man may display in his private conduct and personal relations. Personal or private courage is totally distinct from that higher and nobler courage which prŏmpts the pātriot to ōffer himself a voluntary săcrifice to his country's good.

4. Apprehensions of the imputation of the want of firmness sometimes impel us to perform rash and inconsiderate acts. It is the greatest coŭrage to be able to bear the imputation of the want of courage. But pride, vanity, ēgotism, so unamiable and offensive in private life, are vices which partake of the character of crimes, in the conduct of public affairs. The unfortunate victim of these passions can not see beyŏnd the little, petty, contemptible circle of his own personal ĭnterĕsts. All his thoughts are withdrawn from his country, and concentrated on his consistency, his firmness, himself.

5. The high, the exalted, the sublime emotions of a pātriotism, which, sōaring toward heaven, rises far above all mean, low, or selfish things, and is absorbed by one soul-transpōrting thought of the good and the glōry of one's country, are never felt in his impenetrable bosom. That patriotism, which, cătching its inspirations from the immortal Gōd, and leaving at an immĕasurable distance below all lesser, grŏveling, personal interests and feelings, animates and prŏmpts to deeds of self-săcrifice, of valor,

of devotion, and of death itself,—that is public virtue; that is the noblèst, the sublimest, of all public virtues.

H. CLAY.

CRITICISM.

WHOEVER thinks a faultlèss piece to see
Thinks what ne'er was, nor is, nor e'er shall be.
In every work regard the writer's end,
Since none can compass mōre than they intend ;
And, if the means be just, the conduct true,
Applaüse, in spite of trivial faults, is due.
As men of breeding; sometimes men of wit,
To avoid great errors must the less commit;
Neglect the rules each verbal critic lays;
For not to know some trifles, is a praise.
Mōst critics, fond of some subservient art,
Still make the whōle depend upon a part:
They talk of principles, but notions prize;
And all to one loved folly sacrifice.

POPE.

THE REVOLUTIONARY ALARM.

DARKNESS closed upon the country and upon the town, but it was no night for sleep. Heralds on swift relāys of horses transmitted the war-message from hand to hand, till village repeated it to village; the sea to the backwoods; the plains to the highlands; and it was never suffered to droop; till it had been bōrne North, and South, and East, and West, throughout the land.

2. It spread over the bays that receive the Saco and the Penobscot. Its loud reveille broke the rest of the trappers of New Hampshire, and ringing like bugle-notes from peak to peak, overlēapt the Green Mountains, swept onward to Mŏntreäl, and descended the ocean river, till the responses were echoed from the cliffs of Quebec. The hills along the Hudson told to one another the tale.

3. As the summons hŭrried to the South, it was one day at New York; in one mōre at Phĭlădĕlphĭä; the next it lighted a watchfire at Baltimore; thence it waked an answer at Annapolis. Crossing the Potomac near Mount Vernon, it was sent forward without a halt to Williamsburg. It traversed the Dismal Swamp to Nansemond, ălŏng the route of the first emigrants to North Carolina. It moved onwards and still onwards through boundlèss groves of evergreen to Newbern and to Wilmington.

4. "For Gŏd's sake forward it by night and by day," wrote Cornēliŭs Harnett, by the express which sped for Brunswick. Pātriots of South Carolina caught up its tones at the border and despatched it to Charleston, and through pines and palmetos and mŏss-clad live oaks, further to the South, till it resounded among the New England settlements beyond the Savannah.

5. The Blue ridge took up the voice and made it heard from one end to the other of the valley of Virginiä. The Al'leghānies, as they listened, opened their barriers that the "loud call" might pass through to the hardy riflemen on the Holston, the Watauga, and the French Broad. Ever renewing its strength, powerful enough even to create a commonwealth, it breathed its inspiring word to the first settlers of Kentucky; so that hunters who made their halt in the machlèss valley of the Elkhorn, commemorated the 19th day of April, 1776, by naming their encampment *Lexington*.

6. With one impulse the colonies sprung to arms; with one spirit they pledged themselves to each other "to be ready for the extreme event." With one heart the continent cried, "LIBERTY OR DEATH."

BANCROFT.

SHERIDAN'S RIDE.

1. Up from the South at break of dāy,
   Bringing to Winchester fresh dismāy,
   The affrighted air with a shudder bōre,
   Like a herald in haste, to the chieftain's dōor,
   The terrible grumble, and rumble, and rōar,
   Telling the battle was on once mōre,
   And Sheridan—twenty miles ăwāy.

2. And wider still those billōws of war
   Thundered ălŏng the horĭ'zon's bar,
   And louder yĕt into Winchester rolled
   The rōar of that red sea uncontrolled,
   Making the blood of the listener cold
   As he thought of the stake in that fiery fray,
   And Sheridan—twenty miles away.

3. But there is a rōad from Winchester town,
   A good, broad highway leading down;
   And there, through the flush of the morning light,
   A steed, as black as the steeds of night,
   Was seen to pass as with eagle flight—
   As if he knew the terrible need,
   He stretched away with the utmost speed;
   Hills rose and fell—but his heart was gay,
   With Sheridan fifteen miles away.

4. Still sprung from these swift hoofs, thundering South,
   The dust, like the smoke from the cannon's mouth,
   Or the trail of a comet sweeping faster and faster,
   Foreboding to foemen the doom of disaster;
   The heart of the steed and the heart of the master
   Were beating like prisoners assaulting their walls,
   Impatient to be where the battle-field calls;
   Every nerve of the charger was strained to full play,
   With Sheridan only ten miles away.

5. Under his spurning feet, the rōad
   Like an ărrōwy Al'pĭne river flōwed,

And the landscape sped away behind
Like an ocean flying before the wind;
And the steed, like a bark fed with furnace ire,
Swept on with his wild eyes full of fire.
But, lo! he is nearing his heart's desire—
He is snuffing the smoke of the roaring fray,
With Sheridan only five miles away.

6. The first that the General saw were the groups
Of stragglers, and then the retreating troops;—
What was done—what to do—a glance told him both,
Then striking his spurs with a terrible oath,
He dashed down the line 'mid a storm of huzzahs,
And the wave of retreat checked its course there because
The sight of the master compelled it to pause.
With foam and with dust the black charger was gray;
By the flash of his eye, and his red nostril's play,
He seemed to the whole great army to say,
"*I have brought you Sheridan all the way
From Winchester down to save the day!*"

7. Hurrah, hurrah for Sheridan!
Hurrah, hurrah for horse and man!
And when their statues are placed on high
Under the dome of the Union sky,—
The American soldier's Temple of Fame,—
There, with the glorious General's name,
Be it said in letters both bold and bright:
"*Here is the steed that saved the day
By carrying Sheridan into the fight
From Winchester—twenty miles away!*"

T. B. REED.

THE RAVEN.

1.

ONCE upon a midnight dreary, while I pondered, weak and weary,
Over many a quaint and curious volume of forgotten lore—

While I nodded, nearly napping, suddenly there came a tapping,
As of some one gently rapping, rapping at my chămber-door.
" 'Tis some visitor," I muttered, "tapping at my chamber-door—
   Only this, and nothing more."

2.

Ah, distinctly I remember, it was in the bleak December,
And each separate dying ember wrought its ghost upon the floor,
Eagerly I wished the mŏrrōw: vainly I had sought to bŏrrōw
From my books surcease of sŏrrōw—sorrow for the lŏst Lenōre—
For the rare and rādiant maiden whom the āngels name Lenore—
   Namelĕss here forĕvermōre.

3.

And the silken, sad, uncertain rustling of each purple curtain,
Thrilled me—filled me with fantastic terrors never felt before;
So that now, to still the beating of my heart, I **stood** repeating,
" 'Tis some visitor entreating entrance at my chămber-dōor,—
Some late visitor entreating entrance at my chamber-door—
   That it **is**, and nothing more."

4.

Presently my soul grew **strŏnger**: hesitating then no lŏnger,
"Sir," said I, "or Madam, **truly your** forgivenĕss I implōre;
But the fact is, I was napping, and **so gently you** came rapping,
And **so** faintly you came tapping, **tapping at my** chămber-door,
That I scarce was sure I heard you"—here I opened wide the door,—
   Darkness there, and nothing more.

5.

Deep into that darkness peering, lŏng I stood there wondering, fearing,
Doubting, dreaming dreams no mortals ever dared to dream before;
But the silence was unbroken, and the stillness gave no token,
And the ōnly word there spoken was the whispered word "Lenore!"
This *I* whisper'd, and an echo murmured back the word, "Lenore!"
   Merely this, and nothing more.

6.

Back into the chămber turning, all my soul within me burning,
Soon again I heard a tapping, someting louder than before.
"Surely," said I, "surely that is something at my windōw-lattice;
Let me see then what thereat is, and this mystery explōre,—
Let my heart be still a moment, and this mystery explore;—
   'Tis the wind, and nothing mōre."

7.

Open here I flung the shutter, when, with many a flirt and flutter,
In there stepped a stately raven of the stately days of yōre.
Not the least obēisance made he; not a minute stopp'd or stay'd he;
But, with mien of lord or lady, perch'd above my chămber-door,—
Perch'd upon a bust of Pallas, just above my chamber-door—
   Perch'd, and sat, and nothing more.

8.

Then this ebony bird beguiling my sad fancy into smiling,
By the grave and stern decōrum of the countenance it wōre,

"Though thy crest be shōrn and shaven, thou," I said, "art sure no craven;
Ghastly, grim, and āncient raven, wandering from the nightly shōre,
Tell me what thy lordly name is on the Night's Plutonian shore ?"
    Quoth the raven, "Nevermore!"

9.

Much I marveled this ungainly fowl to hear discōurse so plainly,
Though its answer little meaning—little relevancy bōre;
For we can not help agreeing that no living human being
Ever yĕt was blessed with seeing bird above his chămber-door—
Bird or beast upon the sculptured bust above his chamber-door,
    With such name as "Nevermore!"

10.

But the raven sitting lonely on the placid bust, spoke ōnly
That one word, as if his soul in that one word he did outpōur.
Nothing further then he utter'd—not a feather then he flutter'd—
Till I scarcely more than mutter'd, "Other friends have flown before—
On the mŏrrōw *he* will leave me, as my hopes have flown before."
    Then the bird said, "Nevermore!"

11.

Startled at the stillness, broken by reply so aptly spoken,
"Doubtless," said I, "what it utters is its only stock and stōre,
Caught from some unhappy master, whom unmerciful disaster
Follōw'd fast and follow'd faster, till his sŏngs one burden bōre,—
Till the dirges of his hope that melancholy burden bore,
    Of—"Never—nevermore!"

12.

But the raven still beguiling all my sad soul into smiling,
Straight I wheeled a cushioned seat in front of bird, and bust, and door,
Then, upon the velvet sinking, I betook myself to linking
Fancy unto fancy, thinking what this ominous bird of yŏre—
What this grim, ungainly, ghastly, gaunt, and ominous bird of yore
  Meant in croaking "Nevermore!"

13.

This I sat engaged in guessing, but no syllable expressing
To the fowl, whose fiery eyes now burned into my bosom's cōre;
This and mōre I sat divīning, with my head at ease reclining
On the cushion's velvet lining that the lamp-light glōated ō'ĕr,
But whose velvet viölet lining, with the lamp-light gloating o'er,
  She shall press—ah! nevermore!

14.

Then methought the air grew denser, perfumed from an unseen censer
Swung by seraphim, whose foot-falls tinkled on the tufted floor.
"Wretch," I cried, "thy Gŏd hath lent thee—by these ăngels he hath sent thee
Respite—respite and nepenthe from thy memories of Lenore!
Quaff, oh, quaff this kind nepenthe, and forgĕt this lŏst Lenore!"
  Quoth the raven, "Nevermore!"

15.

"Prophet!" said I, "thing of evil!—prophet still, if bird or devil!
Whether tempter sent, or whether tempest tŏss'd thee here ăshōre,

Desolate, yĕt all undaunted, on this desert land enchanted—
On this home by Hŏrror haunted—tell me truly, I implōre—
Is there—*is* there balm in Gilead?—tell me—tell me, I implore!"
   Quoth the raven, "Nevermore!"

16.

"Prophet!" said I, "thing of evil!—prophet still, if **bird or devil!**
By that heaven **that bends** above us—by that Gŏd we both ădōre,
Tell this soul, with sŏrrōw laden, if, within the distant Aidenn,
It shall clasp a **sainted maiden, whom the ängels name Lenōre;**
Clasp a **rare and** rādient maiden, **whom the angels name** Lenore—
   Quoth the raven, "Nevermore!"

17.

"Be that word our sign of **parting, bird or fiend!**" I shrieked, upstarting—
"Gĕt thee back **into the tempest and the Night's Plutonian** shōre!
Leave no black plume as a token of that lie thy soul hath spoken!
Leave **my** loneliness unbroken!—quit **the bust above my** dōor!
Take thy bĕak **from out my** heart, and take thy form **from** ŏff my door!"
   Quoth the raven, "Nevermore!"

18.

And the **raven,** never flitting, still **is** sitting, **still** is sitting
On the pallid **bust** of Pallas, just above my chamber-door;
And his **eyes** have all the seeming of a dĕmon's that is dreaming,
And the lamp-light o'er him streaming throws his shădōw on the floor;

And my soul from out that shadow that lies flōating on the
floor
   Shall be lifted—NEVERMORE!
          EDGAR A. POE.

THE BELLS.

 HEAR the sledges with the bells—
  Silver bells—
What a world or mĕrriment their melody foretells!
 How they tinkle, tinkle, tinkle,
  In the icy air of night!
 While the stars that oversprinkle
 All the heavens, seem to twinkle
  With a crẙstalline delight;
 Keeping time, time, time,
 In a sort of Runic rhyme,
To the tintinnabulation that so musically wells
 From the bells, bells, bells, bells,
  Bells, bells, bells—
From the jingling and the tinkling of the bells.

2.

 Hear the mĕllōw wedding-bells,
  Gōlden bells!
What a world of happiness their harmony foretells!
 Through the balmy air of night
 How they ring out their delight!
 From the mōlten-gōlden nōtes,
  And all in tune,
 What a liquid ditty flōats
To the turtle-dove that listens, while she glōats
  On the moon!
 Oh, from out the sounding cells,
What a gush of euphony voluminously wells!
  How it swells!
  How it dwells
 On the Future! how it tells
 Of the rapture that impels

SELECTIONS FOR PRACTICE.

To the swinging and the ringing
Of the bells, bells, bells—
Of the bells, bells, bells, bells,
Bells, bells, bells—
To the rhyming and the chiming of the bells!

3

Hear the loud alarum bells—
Brazen bells!
What a tale of tĕrror, now, their turbulency tells!
In the startled ear of night
How they scream out their affright!
Too much hŏrrified to speak,
They can only shriek, shriek,
Out of tune,
In a clamorous appealing to the mercy of the fire,
In a mad expostulation with the dĕaf and frantic fire
Leaping higher, higher, higher,
With a desperate desire
And a resolute endeavor,
Now—now to sit or never,
By the side of the pale-faced moon.
Oh, the bells, bells, bells!
What a tale their terror tells
Of despair!
How they clang, and clash, and rŏar!
What a hŏrror they outpŏur
On the bosom of the palpitating air!
Yĕt the air, it fully knows,
By the twanging
And the clanging,
How the dānger ebbs and flows;
Yet the ear distinctly tells
In the jangling
And the wrangling,
How the danger sinks and swells,
By the sinking or the swelling in the anger of the bells—
Of the bells—

Of the bells, bells, bells, bells,
  Bells, bells, bells—
In the clamor and the clangor of the bells!

4.

Hear the tŏlling of the bells—
  Iron bells!
What a world of solemn thought their monody compels!
  In the silence of the night,
  How we shiver with affright
  At the mĕl'ancholy menace of their tōne!
  For every sound that flōats
  From the rust within their thrōats
    Is a grōan.
  And the people—ah, the people—
  They that dwell up in the steeple,
    All ălōne,
  And who tŏlling, tŏlling, tŏlling,
    In that muffled monotone,
  Feel a glōry in so rōlling
    On the human heart a stone—
  They are nĕither man nor woman—
  They are neither brute nor human—
    They are Ghouls:
  And their king it is who tŏlls;
  And he rōlls, rolls, rolls, rolls,
    A pæan from the bells!
  And his mĕrry bosom swells
    With the pæan of the bells!
  And he dances and he yells;
    Keeping time, time, time,
    In a sort of Runic rhyme,
    To the pæan of the bells—
      Of the bells:
    Keeping time, time, time,
    In a sort of Runic rhyme,
      To the throbbing of the bells—
    Of the bells, bells, bells,
      To the sobbing of the bells;

Keeping time, time, time,
　As he knells, knells, knells,
In a happy Runic rhyme,
　To the rōlling of the bells—
Of the bells, bells, bells—
　To the tōlling of the bells,
Of the bells, bells, bells, bells,
　Bells, bells, bells—
To the mōaning and the grōaning of the bells.

　　　　　　　　　　EDGAR A. POE.

The preceding pieces—"Sheridan's Ride," "The Raven," and "The Bells," are the three most popular in our language, either for private exercise or public declamation. Indeed, any one who can speak them *well* will have little difficulty with ordinary compositions.

CHRISTMAS.

RING out, wild bells, to the wild sky,
　The flying cloud, the frosty light,
　The year is dying with the night;
Ring out, wild bells, and let him die.

Ring out the old, ring in the new,
　Ring happy bells across the snow;
　The year is going, let it go;
Ring out the false, ring in the true.

Ring out the grief that saps the mind,
　For those that here we see no more;
　Ring out the feud of rich and poor,
Ring in redress for all mankind.

Ring out a slowly, dying cause,
　And ancient forms of petty strife;
　Ring in the nobler modes of life,
With sweeter manners, purer laws.

Ring out the want, the woe, the sin,
   The faithless coldness of the times;
Ring out, ring out my mournful rhymes,
But ring the fuller minstrel in.

Ring out false pride in place and blood,
   The civic slander and the spite;
Ring in the love of truth and right,
Ring in the common love of good.

Ring out old shapes of foul disease,
   Ring out the narrowing lust of gold;
Ring out the thousand woes of old,
Ring in the thousand years of peace.

Ring in the valiant man and free,
   The larger heart, the kindlier hand;
Ring out the darkness of the land,
Ring in the CHRIST that is to be.

TENNYSON.

THE TOMAHAWK SUBMISSIVE TO ELOQUENCE.

1. TWENTY tomahawks were raised; twenty ărrŏws drawn to their head. Yĕt stood Harold stern and collected, at bay—parleying ōnly with his swōrd. He waved his arm. Smitten with a sense of their cow′ardĭce, perhaps, or by his great dignity, mōre awful for his vĕry youth, their wĕapons dropped, and their countenances were uplifted upon him, less in hatred than in wonder.

2. The old men găthered about him: he leaned upon his saber. Their eyes shōne with admiration: such heroic depōrtment, in one so young—a boy! so intrĕpid! so prŏmpt! so graceful! so eloquent, too!—for, knowing the effect of eloquence, and feeling the lŏftiness of his own nature, the innocence of his own heart, the character of the Indians for hŏspităl′ity, and their veneration for his

blood, Harold dealt out the thunder of his strength to these rude barbarians of the wilderness, till they, young and old, gathering nearer and nearer in their devotion, threw down their weapons at his feet, and formed a rampart of locked arms and hearts about him, through which his eloquence thrilled and lightened like electricity. The old greeted him with a lofty step, as the pātriarch welcomes his boy from the triumph of far-ŏff battle; and the young clave to him and clung to him, and shouted in their self-abandonment, like brothers round a conquering brother.

3. "Warriors!" he said, "Brethren!"—(their tomahawks were brandished sīmultā'neously, at the sound of his terrible voice, as if preparing for the onset). His tones grew deeper, and less threatening. "Brothers! let us talk together of Logan! Ye who have known him, ye agèd men! bear ye testimony to the deeds of his strength. Who was like him? Who could resist him? Who may ăbīde the hŭrricane in its volley? Who may withstand the winds that uproot the great trees of the mountain? Let him be the foe of Logan. Thrice in one day hath he given battle. Thrice in one day hath he come back victorious. Who may bear up against the strong man—the man of war? Let them that are young, hear me. Let them follōw the cōurse of Logan. He goes in clouds and whirlwind—in the fire and in the smoke. Let them follow him. Warriors! Logan was the father of Harold!" They fell back in astonishment, but they believed him; for Harold's word was unquestioned, undoubted evidence, to them that knew him. NEAL.

CHAPTER VIII.

RULES OF ORDER.

ALL persons who participate in public meetings or debating societies, should make themselves acquainted with the established methods for conducting them. Without a strict adherence to certain recognized rules, it is impossible to avoid confusion and unprofitable wordy controversy. Referring the reader who desires to to be familiar with parliamentary usages in all their applications to "Cushing's Manual," "The American Debater," "The Normal Debater," and similar works, the chapter on this subject will be limited to the necessary rules for managing ordinary Lyceums and debating clubs; and as the Lyceum department of the Hygeio-Therapeutic College has been in existence for more than twenty years, and has simplified its organization to a good, if not the best, working condition, its constitution and by-laws will be presented as a chart or guide for others. This Lyceum has also an uncommon, if not peculiar, feature, which I would strongly commend to all Lyceums whose members are not accomplished speakers. It devotes one whole evening to the discussion of a question agreed on, and another evening to criticisms, readings, essays, and declamations, and so alternately. But, whether this last-named feature is adopted or not; its constitution and by-laws are equally applicable.

CONSTITUTION.

ARTICLE 1.—NAME.

This Association **shall be** entitled, The Hygeio-Therapeutic College Lyceum.

ARTICLE 2.—OBJECTS.

The objects of this Lyceum are, the mutual improvement of its members, and the investigation, in the spirit of candor and truth-seeking, **of** all problems that **concern** the **welfare of human beings.**

ARTICLE 3.—MEMBERSHIP.

Any person may become a member of this Lyceum, on receiving the affirmative vote of two-thirds of the members present at any regular **meeting, and signing** this Constitution.

ARTICLE 4.—EXPULSION.

Any member of this **Lyceum** may be expelled **for** grossly improper conduct, by a **vote** of two-thirds **of the members present** at any regular meeting.

ARTICLE 5.—OFFICERS.

The officers of this Lyceum shall consist of a President, Secretary, **and** Treasurer, who shall exercise their respective duties **for one** week, and until others are chosen **to** succeed them.*

ARTICLE 6.—AMENDMENTS.

This Constitution may be amended at any time by a **vote of** two-thirds of the members present at any regular

---

* A Corresponding Secretary should be elected when the proceedings of the **Society** require letter writing and the circulation of documents.

meeting, provided that notice has been given of the proposed amendment at a preceding regular meeting.

BY-LAWS.

1. MEETINGS.

The Lyceum shall meet in the Lecture Hall of the Hygeio-Therapeutic College, on Monday and Wedsesday evenings, at seven o'clock, and adjourn at nine o'clock.

2. QUORUM.

Five members shall constitute a quorum for the transaction of business. Any number of members less than a quorum may adjourn to the time of the next regular meeting.

3. ORDER OF BUSINESS.

The order of business on Monday evenings shall be:

a. Reading, correction, and adoption of the minutes.
b. Reception of new members.
c. Discussion of the question.
d. Adjournment.

On Wednesday evenings the order of business shall be:

a. Reception of new members.
b. Report of the critic.
c. Criticisms of the critic.
d. Readings, essays, and declamations.
e. Selection of question for debate.
f. Appointments.
g. Unfinished business.
h. New business.
i. Adjournment.

4. DUTIES OF OFFICERS.

The President shall occupy the chair, maintain the order of proceedings, decide all questions of parliamentary usage subject to appeal to the house, appoint all committees, critics, and leading disputants not otherwise provided for, give the casting vote in cases of a tie, and have charge of the books and papers of the Lyceum. The Secretary shall record its proceedings at each meeting, and report the same to the meetings on Monday evenings. The Treasurer shall have charge of the moneys and properties of the Lyceum.

5. APPOINTEES.

On each Wednesday evening a critic, reader, essayist, and declaimer shall be appointed for the ensuing Wednesday evening, and two leading disputants for the discussion on the ensuing Monday evening.*

6. SELECTION OF QUESTION.

The subject for debate shall be selected by a majority vote. Any member may propose, orally or in writing, a question or resolution for discussion.

7. CRITICISMS.

It shall be the duty of the critic to notice all errors in manner, gesture, pronunciation, and grammar, of the preceding meetings, and report the same. After the report of the critic is made, it shall be the privilege of any member to criticise the criticisms of the critic.

---

* When a Lyceum (as in this case) is composed of ladies and gentlemen, it is proper, when practicable, to appoint a lady to open the debate on one side, and a gentleman on the other.

Committees of more than one should be composed of both sexes.

8. LIMITATION OF SPEAKERS.

The leading disputants shall each be entitled to **ten minutes to open, and five minutes to close the debate.** All other speakers shall be **limited to five minutes.** The Lyceum may, at any time, by majority vote, extend the time of any speaker, but not exceeding five minutes.

9. ORDER OF DEBATE.

The affirmative and negative shall be represented alternately from the commencement to the close of the discussion. After the leading disputants have opened the debate, the members shall proceed with the discussion *pro* and *con*, in the order of their names on the book of the Secretary, unless one declines speaking, when the next in order shall be called. If no one offers to controvert the last speaker, another speech on the same side is in order. When all the members who desire to speak have been called, voluntary speakers, *pro* and *con*, may be called for; and if more than one rises to speak, the President shall decide, without appeal or debate, who is entitled to the floor. No one shall be permitted to speak twice until all have spoken who desire to do so, unless by unanimous consent.

POINTS OF ORDER.

All points of order, on being distinctly stated, shall be decided without debate. If the decision of the President is appealed from, the motion, "Shall the decision of the Chair be sustained?" shall be put and decided by a majority vote.

11. MANNER OF VOTING.

Voting may be done by ayes and noes, or by raising the hand, as the Chair shall determine. When the vote is doubtful or disputed, any member may call for a division of the house, when the vote shall be taken by rising or the uplifted hand, the President directing the Secretary to count the ayes and noes.

12. SUSPENSIONS.

Any by-law may be suspended for the evening by a vote of two-thirds of the members present; or it may be suspended indefinitely by unanimous consent.

13. AMENDMENTS.

These by-laws may be amended at any regular meeting of the Lyceum by a vote of two-thirds of the members present; or by a majority vote after one week's notice has been given.

PARLIAMENTARY USAGES.

1. *Motions.*—No motion can be entertained until seconded. When a motion is made and seconded, the President should rise, state the question fully and clearly, and ask if the house is ready for the question. If no one offers to speak, the motion should be put to vote, the result announced, and the Secretary directed to record it.

2. *Motions to Reconsider.*—A motion to reconsider cannot be entertained unless made and seconded by persons who voted with the majority, except in the case of an equal division, when it must be made by one who voted in the negative. No motion to reconsider is in order after the proposition or action has passed out of the possession of the house, or recorded and approved in the minutes.

3. *Motions to Expunge.*—Motions to expunge or rescind any resolution or vote of the house, require unanimous consent.

4. *Motions not Debatable.*—The previous or main question, points of order, motions to reconsider, to adjourn, and to lie on the table, are not debatable; nor are appeals from the decision of the Chair. But when two or more members make an appeal, the President may give his reasons for the decision, and the question may then be debated. In case of a tie vote, the President may give the casting vote in favor of his decision.

5. *The Previous Question.*—The previous question shall not be entertained unless the motion is seconded by three members. If the question is decided affirmatively, and amendments are pending, the vote should be taken first on the amendments in order, and then on the main question. All incidental questions arising after the previous question has been moved, must be decided without debate. When the previous question has been moved and seconded, it cannot be withdrawn without the consent of a majority; nor can it be suspended by any motion except that to adjourn.

6. *Amendments.*—An amendment to a pending motion is always in order; and so is an amendment to an amendment; but an amendment to an amendment cannot be amended. After the discussion the vote is to be taken first on the amendment to the amendment, then on the amendment, and lastly on the main question.

7. *Privileged Questions.*—Privileged questions are those which take precedence of the business regularly before the house. They are:

(*a.*) To adjourn.
(*b.*) For the previous question.
(*c.*) For postponement.
(*d.*) For commitment.
(*e.*) For amendment.

(*f.*) To lie on the table.

A motion for postponement precludes commitment, and a motion for commitment precludes amendment.

8. *Personalities.*—The President may speak in his place to matters of order, or state facts which the members have occasion for. When he rises to speak the member occupying the floor should resume his seat. When a member is speaking, no conversation nor whispering should be indulged in, nor should any one pass between the speaker and the presiding officer. The decision of the President should always be submitted to quietly unless appealed from. A member decided to be out of order loses his right to the floor, without the unanimous consent of the house. No member when speaking should be interrupted, except by a call to order, or a proffer to explain. Members in debate should not refer to the other by name, but as the member who preceded me, last up, on the right, on the left, who opened the debate, etc. No member can be allowed to read an argument, or a paper pertaining to the discussion without unanimous consent. No member can address the house while sitting without unanimous consent. Any member rising to speak should address the President, and not proceed to speak until the President recognizes his right to the floor by announcing his name. When two or more members arise to speak at the same time, the President shall decide who is entitled to the floor by announcing his name, or designating him in some other manner. The motives of members are never to be questioned.

9. *Appeals.*—Any member may appeal from any decision of the Chair; but the member appealing must reduce his appeal to writing, and hand it to the Secretary. The President shall then state the question, and call for

a vote on the question, "Shall the decision of the Chair be sustained?"

10. *Explanations.*—No explanation can be made while a member is speaking without the consent of the speaker; but if the speaker yields the floor for an explanation, he cannot resume it again without unanimous consent. Members who obtain leave to explain must confine their remarks to the matters to be explained.

Committees.—In legislative bodies, committees are of two kinds, select or special, and standing or permanent. In Lyceums all committees are of the former kind. Their duties are to consider any subject or proposition referred to them, and report the same to the next meeting, or at any time designated. They may report in full or ask to be discharged, or report progress and ask leave to be continued. Their report may be considered and disposed of as a whole, or in sections or parts, when the subject is susceptible of such division. In the latter case each section may be approved, rejected, or amended, and then the final vote taken, whether it shall be adopted or rejected as a whole. The first person named on a committee of several usually acts as chairman.

11. *Postponements.*—These may be for the time, or indefinitely. When different times are mentioned the question should be taken on the most distant time first. The motion to postpone indefinitely cannot be amended, nor superseded by any other motion; but if decided negatively, a motion to amend or commit will be in order.

12. *Adjournment.*—A motion to adjourn is not in order when a member is speaking, nor when a vote is being taken on any question. When a motion to adjourn has been negatived, it cannot be renewed until some other

proposition has been presented, or business of some kind transacted. A motion to adjourn cannot be amended by adding to it a definite time or place; this must be previously decided on its own merits. A motion to adjourn to a particular time and place is debatable so far as the time and place are concerned. When desiring to suspend business temporarily, an adjournment for the time is in order, after which the business may be resumed on a simple motion to do so. When an adjournment has been voted during the consideration of any question, that question will be first put in order among the unfinished business, but not the first business in order at the next meeting.

CHAPTER IX.

DEBATABLE SUBJECTS.

The following questions are submitted for emergencies —when for want of time or for some other reason, the Lyceum is unable to agree on any question presented by the members.

Can a law of nature be suspended?
Ought the elective franchise to be **extended to** woman?
Are the sexes equal in mentality?
Is the female organization naturally more frail than that of the male?
Should public libraries be opened on Sunday?
Are schools or churches the greater benefit to society?
Is there more pleasure in pursuit than in possession?
Should immigration to this country be restricted?
Should eight hours be recognized as a legal day's work?

Resolved, that the veto power of the President be repealed.

Resolved, that all punishment should be limited to the reformation of the criminal.

Resolved, that capital punishment should be abolished.

Resolved, that all laws relating to interest on money should be repealed.

DEBATABLE SUBJECTS.

Resolved, that interest on money should be limited to the profits of productive industry.

Resolved, that all laws for the collection of debts should be repealed.

Resolved, that the rate of taxation should have reference to the property of the person taxed.

Resolved, that we suffer more from imaginary than from real evils.

Has the human race descended from a single pair?

Resolved, that conscience is an infallible rule for action.

Is the medical profession more useful than injurious?

Should national holidays be abolished?

Should the sexes be educated in the same schools?

Does geology harmonize with the Bible?

Should education be compulsory?

Is woman physiologically the "weaker vessel?"

Does civilization progress more rapidly than the churches?

Is the theory of Darwin, as to the "descent of man," sustained by scientific data?

Is the doctrine of "Evolution" taught in the Bible?

Should white and colored children attend the same school?

Should society permit the existence of dram-shops?

Are alcoholic drinks a greater evil than tobacco?

Is the dietetic character of man frugivorous?

Is common salt useful as a condiment?

Should the property of churches be taxed?

Do labor-saving inventions benefit the laboring classes?

Are trades-unions justifiable?

Do great crises produce great men?

Ought old bachelors to be subject to civil disabilities?

Should monopolies in trade be allowed?

Ought ministers of the Gospel to engage in party politics?
Ought there to be a law of international copyright?
Is universal suffrage expedient?
Can the immortality of the soul be proved from the light of nature?
Do riches develop character better than poverty?
Is Roman Catholicism compatible with free institutions?
Ought imprisonment for debt to be abolished?
Is infidelity on the increase?
Is Phrenology a true science?
Is the assassination of tyrants justifiable?
Ought lotteries to be tolerated?
Are religious fairs justifiable?
Are ghosts the spirits of the departed?
Is youth a more happy period of life than old age?
Do preachers exercise a greater influence on the character of the young than teachers?
Are lawyers more beneficial than injurious to society?
Are women more revengeful than men?
Ought persons to marry who differ radically in religious opinions?
Are "all men created equal?"
Is the doctrine of original sin taught in the Bible?
Is morality separable from religion?
Does morality improve as civilization advances?
Is a Republic the best form of government?
Is the character of a nation affected by its climate?
Ought witnesses to be held as prisoners?
Is a declaration of war ever justifiable?
Resolved, that Satan is the hero of "Paradise Lost."
Is there such a quality as disinterestedness.
Ought patent-rights to be granted?

DEBATABLE SUBJECTS.

Does wealth exert more influence than knowledge?
Are banks more beneficial than injurious to a community?
Is there any real danger of over-population?
Are national celebrations beneficial?
Are persons accountable for their opinions?
Is man a free agent?
Are tea and coffee, as beverages, injurious?
Is it hygienic to drink at meals?
Is there more happiness than misery in human life?
Should the Bible be introduced into the common schools?
Is a falsehood ever justifiable?
Is the doctrine of non-resistance sound?
Resolved, that differences of character are attributable more to physical than to moral causes.
Is the slanderer a more pernicious character than the flatterer?
Do the phenomena of nature indicate polytheism?
Are ideas innate?
Ought emulation in schools to be encouraged?
Is corporeal punishment in schools justifiable?
Is rotation in office a correct principle?
Is it ever right to marry for money?
Is it expedient to wear mourning apparel?
Are graveyards expedient?
Would the practice of cremation be beneficial?
Is the miser more selfish than the profligate?
Ought one ever to advocate what he believes to be false?
Does proselytism favor the cause of truth?
Is the drunkard accountable for his conduct while drunk?
Do the Scriptures predict a millenium?

THE INDISPENSABLE HAND-BOOK.

How to Write---How to Talk---How to Behave, and How to Do Business.

COMPLETE IN ONE LARGE VOLUME.

This new work—in four parts—embraces just that practical matter-of-fact information which every one—old and young—ought to have. It will aid in attaining, if it does not insure, "success in life." It contains some 600 pages, elegantly bound, and is divided into four parts, as follows:

How to Write:

AS A MANUAL OF LETTER-WRITING AND COMPOSITION, IS FAR SUPERIOR to the common "Letter-Writers." It teaches the inexperienced how to write Business Letters, Family Letters, **Friendly Letters**, **Love Letters**, **Notes and Cards, and Newspaper Articles**, and how to Correct Proof for the Press. The newspapers have pronounced it "Indispensable."

How to Talk:

NO OTHER BOOK CONTAINS SO MUCH USEFUL INSTRUCTION ON THE subject as this. It teaches how to Speak Correctly, Clearly, Fluently, Forcibly, **Eloquently, and Effectively**, in the Shop, in the Drawing-room; a Chairman's Guide, to conduct Debating Societies and Public Meetings; how to Spell, and how to Pronounce all sorts of Words; with Exercises for **Declamation**. The chapter on "Errors Corrected" is worth the price of the volume to every **young man**. "Worth a dozen grammars."

How to Behave:

THIS IS A MANUAL OF ETIQUETTE, AND IT IS BELIEVED TO BE THE best "MANNERS BOOK" ever written. If **you** desire to know what good manners require, at Home, on the Street, at a Party, **at** Church, at Table, in Conversation, at Places of Amusement, in Traveling, in the Company of Ladies, in Courtship, this book will inform you. It is a standard work on Good Behavior.

How to Do Business:

INDISPENSABLE IN THE COUNTING-ROOM, **IN THE** STORE, IN THE SHOP, on the FARM, for the Clerk, the Apprentice, the Book Agent, and for Business Men. It teaches how to Choose a Pursuit, and how to follow it with success. "It teaches how to get rich honestly," **and** how to use your riches wisely.

How to Write—How to Talk—How to Behave—How **to Do Business, bound** in **one large** handsome **volume**, post-paid, for **$2 25.**

Agents wanted. Address, S. R. WELLS & CO., 737 Broadway, New York.

Works on Hygiene,

By R. T. TRALL, M. D.

Hydropathic Encyclopedia, A System of Hygiene, embracing Outlines of Anatomy—Physiology of the Human Body; Preservation of Health, Dietetics and Cookery; Theory and Practice of Hygienic Treatment; Special Pathology and Therapeutics, including the Nature, Causes, Symptoms, and treatment of all known Diseases; Application of Hydropathy to Midwifery and the Nursery. Nearly 1,000 pages, including a Glossary, Table of Contents, and a complete Index. A Guide to Families and Students, and a Text-Book for Physicians. With 300 illustrations. In one vol. $4.50.

Anatomical and Physiological Plates. These Plates were arranged expressly for Lecturers on Health, Physiology, etc. Six in number, representing the normal position and life-size of all the internal viscera, magnified illustrations of the organs of the special senses, and a view of the nerves, arteries, veins, muscles, etc. Price, fully colored, backed, and mounted on rollers (net) $20, by express.

The Hygienic Hand-Book, Intended as a Practical Guide for the Sick Room, arranged Alphabetically. With Appendix Illustrative of the Hygeo-Therapeutic Movements. $2.00.

Family Gymnasium, Containing the most approved methods of applying Gymnastic, Calisthenic, Kinesipathic and Vocal Exercises to the Development of the Bodily Organs, Invigoration of their Functions, Preservation of Health, and the Cure of Disease and Deformities. $1.75.

Hydropathic Cook-Book, With Recipes for Cooking on Hygienic Principles. An Exposition of the Relations of Food to Health; Chemical Elements and Proximate Constitution of Alimentary Principles: Relative Value of Vegetable and Animal Substances, $1.50.

Digestion and Dyspepsia, A Complete Explanation of the Physiology of the Digestive Processes, with Symptoms and Treatment of Dyspepsia and other Disorders. $1.00.

Mother's Hygienic Hand-Book, For the Normal Development and Training of Women and Children, and the Treatment of their Diseases. $1; fine, $1.25.

The Human Voice; Its Anatomy, Physiology, Pathology, Therapeutics and Training, with Rules of Order for Lyceums. More than 50 Illustrations. Plain, 50 cents; fine, $1.00. In Press.

Water-Cure for the Million. The Processes of Water-Cure Explained. Rules for Bathing, etc., given. Recipes for Cooking, Directions for Home Treatment. 30c.; muslin, 50c.

The True Healing Art, Or Hygienic vs. Drug Medication. A plain, practical view of the question, including an address before Smithsonian Institute. 30 cents; muslin, 50 cents.

The Hygeian Home Cook-Book, Or Palatable Food without Condiments. A Complete Book of Recipes or Directions for Preparing and Cooking all kinds of Healthful Food in a Healthful Manner. 25 cents; in muslin, 50 cents.

The Alcoholic Controversy, A Review of the Physiological Errors of Teetotalism. 50 cts.

Diseases of Throat and Lungs, Including Diphtheria, and their Proper Treatment. 25c.

The Bath. Its History and Uses in Health and Disease, with Twenty Engravings. The best work of the kind. 25 cents; muslin, 50 cents.

The New Health Catechism, Questions and Answers on How to Live. 12mo. Paper, 10 cts.

Sent by mail, by S. R. WELLS & CO., 737 Broadway, New York.

"IT IS AN ILLUSTRATED CYCLOPEDIA."

NEW PHYSIOGNOMY;

OR,

SIGNS OF CHARACTER,

As manifested in Temperament and External Forms, and especially in the Human Face Divine.

BY S. R. WELLS, EDITOR PHRENOLOGICAL JOURNAL.

Large 12mo, 768 pp. With more than 1,000 Engravings.

Illustrating Physiognomy, Anatomy, Physiology, Ethnology, Phrenology, *and* Natural History.

A COMPREHENSIVE, thorough, and practical Work, in which all that is known on the subject treated is Systematized, Explained, Illustrated, and Applied. Physiognomy is here shown to be no mere fanciful speculation, but a consistent and well-considered system of Character-reading, based on the established truths of Physiology and Phrenology, and confirmed by Ethnology, as well as by the peculiarities of individuals. It is no abstraction, but something to be made useful; something to be practiced by everybody and in all places, and made an efficient help in that noblest of all studies—the Study of Man. It is readily understood and as readily applied. The following are some of the leading topics discussed and explained in this great Illustrated work:

Previous Systems given, including those of all ancient and modern writers.

General Principles of Physiognomy, or the Physiological laws on which character-reading is and must be based.

Temperaments.—The Ancient Doctrines—Spurzheim's Description—**The New Classification** now in use here.

Practical Physiognomy.—General Forms of Faces—The Eyes, the Mouth, the Nose, the Chin, the Jaws and Teeth, the Cheeks, the Forehead, the Hair and Beard, the Complexion, the Neck and Ears, the Hands and Feet, the Voice, the Walk, the Laugh, the Mode of Shaking Hands, Dress, etc., with illustrations.

Ethnology.—The Races, including the Caucasian, the North American Indians, the Mongolian, the Malay, and the African, with their numerous subdivisions; also National Types, each illustrated.

Physiognomy Applied.—To Marriage, to the Training of Children, to Personal Improvement, to Business, to Insanity and Idiocy, to Health and Disease, to Classes and Professions, to Personal Improvement, and to Character-Reading generally. Utility of Physiognomy, Self-Improvement.

Animal Types.—Grades of Intelligence, Instinct and Reason—Animal Heads and Animal Types among Men.

Graphomancy.—Character revealed in Hand-writing, with Specimens—Palmistry. "Line of Life" in the human hand.

Character-Reading.—More than a hundred noted Men and Women introduced—What Physiognomy says of them.

The Great Secret.—How to be Healthy and How to be Beautiful—Mental Cosmetics—very interesting, **very** useful.

Aristotle and St. Paul.—A Model Head—Views of **Life**—Illustrative Anecdotes—Detecting a Rogue by his Face.

No one can read this Book without interest, without real profit. "Knowledge is power," and this is emphatically true of a knowledge of men—of human character. He who has it is "master of the situation;" and anybody may have it who will, and find in it the "secret of success" and the road to the largest personal improvement.

Price, in one large Volume, of nearly 800 pages, and more than 1,000 engravings, on toned paper, handsomely bound in embossed muslin, $5; in heavy calf, marbled edges, $8; Turkey morocco, full gilt, $10. Agents may do well to **canvass** for this work. Free by post. Please address, S. R. WELLS & CO., 737 Broadway, New York.

Now Ready, a New and Useful Work for Young People.

WEDLOCK;

OR, THE RIGHT RELATIONS OF THE SEXES—Disclosing the Laws of Conjugal Selection, and showing **Who** May and **Who** May Not Marry. A Scientific Treatise. By SAMUEL R. WELLS. One vol., 12mo, 250 pages; plain **muslin,** price, **$1 50;** in fancy gilt binding, **$2.** Published by the **Author,** at 737 Broadway, New York.

Among **the** subjects **treated** are the following: Marriage **a** Divine Institution; Qualifications for Matrimony; **The** Right Age to Marry; Motives **for** Marrying; Marriages of Consanguinity—of Cousins, when Justifiable; Conjugal Selection—Affinities; Courtship—Long or Short; Duty of Parents; Marriage Customs and Ceremonies of all Nations; Ethics of Marriage; Second Marriages, are they Admissible; Jealousy— Its Cause and Cure; Causes of Separation and Divorce; Celibacy— Ancient and Modern; Polygamy and Pantagamy; Love Signs in the Features, and How to Read Them; Physiognomy; Sensible Love Letters—Examples; The Poet's Wife; The Model Husband and the Model Wife—their Mutual Obligations, Privileges, **and** Duties; **The Poetry** of Love, Courtship, and Marriage—Being **a** Practical Guide to all the Relations of HAPPY WEDLOCK.

Here are some of the contents, compiled from the Index, which give a more definite idea of the scope and objects of the work:

Development and Renewal **of** the Social Affections; Inordinate Affection; Function **of** Adhesiveness and Amativeness; Admiration not Love; Addresses Declined, How to Do It; The Bible on Marriage; Matrimonial Bargains; True Beauty; Celibacy and Health; Celibacy and Crime; Marrying for Money; Facts in Relation to Consanguineous Marriage—when Permissible; Law of Conjugal Selection; **Conjugal Harmony;** Conjugal Resemblances of **Husbands and Wives;** Pleasure **of** Courtship; **Confidence in Love;** Duty of Cheerfulness; **Woman's** Constancy; Laws and Remedy **for Divorce;** Drifting; Economy; Etiquette of Long Engagements; Falling in Love; Forbearance; Whom Great Men Marry; Girls of the Period; Housekeeping; Good Habits Essential; How to Win Love; Honeymoon; The Model Husband; Home, How to Make it Happy; Mutual Help; Conjugal Harmony; Hotel and Club Life; Inhabitiveness; Terrible Effects **of** Morbid Jealousy; Juliet's Confession; **Kisses;** Kate's Proposal; Parental Love, **How to** Win it; Declarations of Love; Not to be Ashamed of it; Romantic Love; Second Love, Is Love Unchangeable? Should Parents Interfere; Love-Letters; **Love** Song; Congratulatory Letter; Little Things; Love's Seasons; Its Philosophy; Early Marriage among the Ancients; Motives for it; International Marriage; Marriage Customs, Marriage Defined; Its Legal Aspects; Marriage Ceremonies in the Episcopal, the Roman, and in the Greek Churches, Jewish and Quaker; Marriage Exhortation; Prayer; Hymns; Ethics of Marriage; Health and Marriage; Hasty Marriages; Marriage Maxims; Morganatic Marriages; Marrying for a Home, for Money, for Love, for Beauty; Right Motive for Marrying; Advice to the Married; Man and Woman Contrasted; Monogamy Defined; Matrimonial Fidelity; Matrimonial Politeness; Legal Rights of Married Women; The Mormon System; Man's Requirements; The Maiden's Choice; Letters of Napoleon; When to Pop the Question; Pantagamy at Oneida Defined; Meddling Relatives; Physical and Mental Soundness; Step-Mothers; The Shakers; Singleness; Sealing; Something to Do; Wedding in Sweden; Temptations of the Unmarried; Hereditary Taints; Temperaments; Trifling; Too Much to Do; May Women Make Love; Lesson for Wives; Wedding Gifts; Wife and I; Yes, How a Lady Said It; Plain Talk with a Young Man; Soliloquy of a Young Lady, and much more, covering the whole ground of Marriage. A beautiful Gift-Book for all seasons.

The book **is** handsomely **printed** and beautifully bound. It was intended more especially **for young people,** but may be read with interest and with profit by those **of every age.** Copies will be sent by post to any address on receipt of **price,** by **S. R.** Wells & Co., 737 Broadway, N. Y.

"GOOD BOOKS FOR ALL."

S. R. WELLS & CO., Publishers, 737 Broadway, New York.

Best Works on these subjects. Each covers ground not covered by others. Copies sent by return post, on receipt of price. Please address as above.

American Phrenological Journal and Life Illustrated. Devoted to Ethnology, Physiology, Phrenology, Physiognomy, Psychology, Biography, Education, Art, Literature, with Measures to Reform, Elevate and Improve Mankind Physically, Mentally and Spiritually. Monthly, $3 a year.

Annuals of Phrenology and Physiognomy. One yearly 12mo volume. Price 25 cents for the current year. For 1865, '66, '67, '68, '69, '70, '71, '72 and '73. The nine containing over 400 pages, many portraits, with articles on "How to Study Phrenology," "Bashfulness, Diffidence, Stammering," "Marriage of Cousins," "Jealousy, Its Cause and Cure," etc. Bound in one vol., $2.

Constitution of Man. Considered in relation to External Objects. By GEORGE COMBE. The only authorized American Edition. Twenty Engravings, and a Portrait of Author. $1.75.

Chart for Recording Development, 10 cents.

Chart of Physiognomy Illustrated for Framing. Map. 25 cents.

Defence of Phrenology. A Vindication of Phrenology against Attacks. The Cerebellum the seat of the reproductive instinct. BOARDMAN. $1.50.

Domestic Life. Marriage Vindicated. Free Love Exposed. By SIZER. 25 cents.

Education. Founded on the Nature of Man. By J. G. SPURZHEIM. Appendix, with the Temperaments. $1.50.

Education and Self-Improvement Complete. Physiology—Animal and Mental; Self-Culture and Perfection of Character; Memory and Intellectual Improvement. In one vol. Muslin, $4.

Expression: Its Anatomy and Philosophy. By SIR CHARLES BELL. Illustrations. Notes by Editor *Phrenological Journal.* Fancy cloth. Fine. 1.50.

How to Read Character. A New Hand-Book of Phrenology and Physiognomy, for Students and Examiners, with a Chart for recording the sizes of the different Organs of the Brain, in the Delineation of Character. 170 Engravings. Latest and best. Paper, $1. Muslin, 1.25.

Memory and Intellectual Improvement, applied to Cultivation of Memory. Very useful. $1.50.

Lectures on Phrenology. By GEORGE COMBE. The Phrenological Mode of Investigation. 1 vol. 12mo, $1.75.

Mental Science. The Philosophy of Phrenology. By WEAVER. $1.50.

Moral Philosophy. By GEORGE COMBE. Or, the Duties of Man considered in his Individual, Domestic and Social Capacities. Latest Ed. $1.75.

Natural Laws of Man. Questions with Answers. A Capital Work. By J. G. SPURZHEIM, M.D. Muslin, 75 cents.

New Physiognomy; or, Signs of Character, as manifested through Temperament and External Forms, and especially in the "Human Face Divine." With 1000 *Illustrations*. In three styles of binding. In muslin, $5; in heavy calf, $8; turkey morocco, full gilt, $10.

Phrenology and the Scriptures. Harmony between Phrenology and Bible. By REV. JOHN PIERPONT. 25 cents.

Phrenological Bust. Showing the latest classification, and exact locations of the Organs of the Brain, for Learners. In this Bust, all the newly-discovered Organs are given. It is divided so as to show each Organ on one side; and all the groups—Social, Executive, Intellectual, and Moral—properly classified, on the other side. Two sizes; largest in box, $2.00; smaller, $1.00. Sent by express.

Phrenological Guide for Students, 25 cents.

Phrenology Proved, Illustrated and Applied. Analysis of the Primary Mental Powers in their Various Degrees of Development, and Location of the Phrenological Organs. $1.75.

Self-Culture and Perfection of Character; Including the Training and Management of Children. $1.50.

Self-Instructor in Phrenology and Physiology. 100 Engravings, Chart for Phrenologists. Pap. 50 cts. mus. 75 cts.

Symbolical Head and Phrenological Map, for Framing. 25 cents.

Wells' New Descriptive Chart for the Use of Examiners, giving a Delineation of Character. Illus. 25 cents.

Your Character from Your Likeness. Inclose stamp for a copy of circular, "Mirror of the Mind."

To Physicians, Lecturers, and Examiners. We have a Cabinet of 40 Casts of Heads, selected from Our Museum, which are sold at $35. Also a set of Phrenological Drawings on canvas, size of life, 40 in number price $40. A set of six Anatomical and Physiological plates, colored and mounted, $20. Another set of twenty, in sheets, plain, $35. Colored and mounted, $60. Skeletons, from $50 to $100. Manikins, $500 to $1000. Portraits in oil from $5 upwards. Woodcuts, $3.50 to $5. Symbolical Heads. Electrotypes, $3 to $5, and $7.50, each.

All Works pertaining to the "SCIENCE OF MAN," including Phrenology, Physiognomy, Ethnology, **Psychology**, Physiology, Anatomy, Hygiene, **Dietetics**, etc., supplied. Enclose stamp for Wholesale Terms.

Works on Physiology and Hygiene.

[It has been said that, a man at Forty Years of Age, is either "a Physician or a Fool." That at this Age, he ought to know how to treat, and take care of himself. These Works are intended to give instruction on "How to Live," and How to avoid Diseases.]

The Science of Health. A new Independent Health Monthly, which teaches the Laws by which Health is preserved, Disease eradicated, and Life prolonged, on Hygienic Principles. Its agencies are: Food, Drink, Air, Exercise, Light, Temperance, Sleep, Rest, Bathing, Clothing, Electricity, Right Social Relations, Mental Influences. It is a first-class Magazine, published at $2 a year.

Anatomical and Physiological PLATES Arranged expressly for Lectures on Health, Physiology, etc. By R. T. Trall, M.D. They are six in number, representing the normal position and life-size of all the internal viscera, magnified illustrations of the organs of the special senses, and a view of the nerves, arteries, veins, muscles, etc. Fully colored, backed, and mounted on rollers. Price for the set, net $20.

The Mother's Hygienic Hand-Book, for the normal development and training of women and children, and the treatment of their diseases. TRALL. $1.

Accidents and Emergencies, What to Do—How to Do it. 25 cents.

Cure of Consumption by Mechanical Movements. By Dr. Wark, 30c.

Children, their Management in Health and Disease. By Dr. Shew. $1.75.

Diseases of the Throat and LUNGS. With Treatment. 25 cents.

Digestion and Dyspepsia. Complete Explanation of the Physiology of the Digestive Processes, Symptoms and Treatment of Dyspepsia and other Disorders of the Digestive Organs. Illus. By Dr. TRALL. New, Muslin, $1.

Philosophy of Electrical Psychology, in 12 Lectures. Dodds. $1.50.

Family Gymnasium. Gymnastic, Calisthenic, Kinesipathic, and Vocal Exercises, Development of the Bodily Organs. By Dr. Trall. Illus. $1.75.

Domestic Practice of Hydropathy. By E. Johnson, M.D. $2.

Hydropathic Cook Book. Recipes for Cooking on Hygienic Principles, $1.50.

Food and Diet. With Dietetical Regimen suited for Disordered States of the Digestive Organs. Dietaries of Principal Metropolitan Establishments for Lunatics, Criminals, Children, the Sick, Paupers, etc. A Scientific Work. By Pereira Edited by Dr. C. A. Lee. $1.75.

Fruits and Farinacea, the PROPER FOOD OF MAN. By Dr. Smith. With Notes by Dr. Trall. $1.75.

Hydropathic Encyclopedia. A System of Hydropathy and Hygiene. Outlines of Anatomy; Physiology of the Human Body; Hygienic Agencies, Preservation of Health; Theory and Practice; Special Pathology, Nature, Causes, Symptoms, and Treatment of all known Diseases. A Guide to Families and Students, and a Text-Book for Physicians. By R. T. Trall, M.D. $4.50. "The most complete work on this subject." *Tribune.*

Hygienic Hand-Book. A Practical Guide for the Sick-Room. Alphabetically arranged. Appendix. Trall. $2.

Family Physician. A Ready Prescriber and Hygienic Adviser. With Reference to the Nature, Causes, Prevention and Treatment of Diseases, Accidents, and Casualties of every kind. With a Glossary and Copious Index. Illustrated. By Joel Shew, M.D. Muslin, $4.

Management of Infancy, Physiological and Moral Treatment. By Andrew Combe, M.D. Muslin, $1.50.

Medical Electricity. Manual for Students, Showing the scientific application to all forms of Acute and Chronic Diseases of the Different combinations of Electricity, Galvanism, Electro-Magnetism, Magneto-Electricity, and Human Magnetism. By Dr. White, $2.

Midwifery and the Diseases of Women. A Descriptive and Practical Work. With the general management of Child-Birth, Nursery, etc. $1.75.

Movement-Cure (Swedish). History and Philosophy of this System of Medical Treatment, with Examples and Directions for their Use in Diseases. *Illustrated.* By Dr. Taylor. $1.75.

Notes on Beauty, Vigor and Development; or, How to Acquire Plumpness of Form, Strength of Limb, and Beauty of Complexion. 12 cents.

Paralysis and other Affections of the Nerves; Cure by Vibratory and Special Movements. Dr. Taylor. $1.

Physiology of Digestion. Considered with relation to the Principles of Dietetics. *Illus.* Combe. 50 cents.

Parents' Guide; or, Human Development through Inherited Tendencies. By Mrs. Hester Pendleton, Second ed., revised and enlarged. 1 vol. 18mo, $1.50.

Philosophy of Mesmerism and Clairvoyance. Six Lectures, with Introduction, 50 cents.

Philosophy of Sacred History, considered in Relation to Human Aliment and the Wines of Scripture. By Sylvester Graham, $3.00.

Philosophy of Water Cure. Development of the true Principles of Health and Longevity. Balbirnie. 50 cents.

Practice of Water Cure. Containing a Detailed account of the various Bathing processes. *Illus.* 50 cents.

Physiology, Animal and Mental: Applied to the Preservation and Restoration of Health, of Body and Power of Mind. *Illus.* Muslin, $1.50.

Principles of Physiology applied to the Preservation of Health and to the Improvement of Physical and Mental Education. By Andrew Combe. $1.75.

Sober and Temperate Life. Discourses and Letters of Cornaro. 50 cts.

Science of Human Life, Lectures On. By Sylvester Graham. With a copious Index and Biographical Sketch of the Author. *Illustrated.* $3.50.

Story of a Stomach. An Egotism. By a Reformed Dyspeptic. Mus. 75 cts.

Tea and Coffee, their Physical, Intellectual, and Moral Effects on the System. By Dr. Alcott. 25 cents.

Teeth; their Structure, Disease and Management, 25 cents.

Tobacco; Its Physical, Intellectual and Moral Effects, 25 cents.

The Alcoholic Controversy. A Review of the *Westminster Review* on Errors of Teetotalism. Trall. 50 cents.

The Bath. Its History and Uses in Health and Disease. By R. T. Trall, M.D. Paper, 25 c.; muslin, 50 cents.

The Hygeian Home Cook Book. Palatable Food without Condiments. 25c.

The Human Feet. Their Shape, Dress and Proper Care, $1.25.

True Healing Art; or, Hygienic vs. Drug Medication. A practical view of the whole question. Pap. 30 c., mus. 50 c.

Water-Cure for the Million. The Processes Explained. Popular Errors Exposed. Hygienic and Drug Medication Contrasted. Rules for Bathing, Dieting, Exercising, etc. Pap., 30 cts., mus. 50 cts.

Water-Cure in Chronic Diseases. Causes, Progress, and Terminations of Various Diseases of the Digestive Organs, Lungs, Nerves and Skin, and their Treatment. By Dr. Gully. $2.

"Special List" of 70 or more Private Medical, Surgical and Anatomical Works, invaluable to those who need them, sent on receipt of stamp, by S. R. Wells, N.Y.

Works for Home Improvement.

This List embraces just such Works as are suited to every member of the family—useful to all, indispensable to those who have not the advantages of a liberal education.

Aims and Aids for Girls and Young Women, on the Duties of Life, Self-Culture, Dress, Beauty, Employment, Duties to Young Men, Marriage, and Happiness. By Weaver. $1.50.

Æsop's Fables. The People's Pictorial Edition. Beautifully Illustrated with nearly Sixty Engravings. Gilt. $1.

Carriage Painter's Illustrated Manual. Art, Science and Mystery of Coach, Carriage, and Car Painting. Fine Gilding, Bronzing, Staining, Varnishing, Polishing, Copying, Lettering, Scrolling, and Ornamenting. By Gardner. $1.00

Chemistry. Application to Physiology, Agriculture, Commerce. Liebig. 50 cents.

Conversion of St. Paul. By Geo. Jarvis Geer, D.D. 12mo. $1.

Footprints of Life; or, Faith and Nature Reconciled. A Poem in Three Parts. The Body, The Soul, The Deity. By Philip Harvey, M.D. $1.25.

Fruit Culture for the Million: A Hand-Book. Being a Guide to the Cultivation and Management of Fruit Trees. How to Propagate them. *Illus.* $1.

Gems of Goldsmith.—The Traveler. The Deserted Village. The Hermit. With notes and a sketch of the Great Author. *Illustrations.* Gilt. $1.

Good Man's Legacy. Rev. Dr. Osgood. 25 cents. **Gospel Among the Animals.** Same. 25 cents.

Hand-Book for Home Improvement: Comprising "How to Write," "How to Talk," "How to Behave," and "How to do Business." One vol. $2.25.

How to Live; Saving and Wasting, or, Domestic Economy made plain. $1.50.

How to Paint. A Complete Compendium of the Art. Designed for the Use of Tradesmen, Mechanics, Merchants, Farmers, and a Guide to the Professional Painter. By Gardner. $1.

Home for All; The Concrete, or Gravel Wall. Octagon. New, Cheap, Superior Mode of Building. $1.50.

Hopes and Helps for the Young of Both Sexes. Formation of Character, Choice of Avocation, Health, Conversation, Cultivation, Social Affection, Courtship, Marriage. Weaver. $1.50.

Library of Mesmerism and Psychology. Philosophy of Mesmerism, Clairvoyance, and Mental Electricity; Fascination, or the Power of Charming; The Macrocosm, or the World of Sense; Electrical Psychology, Doctrine of Impressions; The Science of the Soul, Physiologically and Philosophically. $4.

Life at Home; or, The Family and its Members. A capital work. By William Aikman, D.D. $1.50; gilt, $2.

Life in the West; or, Stories of the Mississippi Valley. Where to buy Public Lands. By N. C. Meeker. $2.

Man and Woman considered in their Relation to each other and to the World. By H. C. Pedder. $1.00.

Man, in Genesis and in Geology; or, the Biblical Account of Man's Creation, tested by Scientific Theories of his Origin and Antiquity. By Joseph P. Thompson, D.D., LL.D. One vol. $1.

Pope's Essay on Man. With Notes. Beautifully Illustrated. Cloth, gilt, beveled boards. Best edition. $1.

A Self-Made Woman; or, Mary Idyl's Trials and Triumphs. Suggestive of a noble and happy life. By Emma M. Buckingham. $1.50.

Oratory—Sacred and Secular; or, the Extemporaneous Speaker. Including Chairman's Guide. $1.50.

Temperance in Congress. Ten Minutes' Speeches—powerful—delivered in the House of Representatives. 25 cts.

The Christian Household. Embracing Husband, Wife, Father, Mother, Child, Brother, Sister. Weaver. $1.

The Emphatic Diaglott; or, the New Testament in Greek and English. Containing the Original Greek Text of the New Testament, with an Interlineary Word-for-Word English Translation. By Wilson. Price $4, extra fine binding, $5.

Right Word in the Right Place. A New Pocket Dictionary Synonyms, Technical Terms, Abbreviations, Foreign Phrases, Writing for the Press, Punctuation, Proof-Reading. 75 cents.

The Temperance Reformation. History from the first Temp. Society in U. S. By Armstrong. $1.50.

The Model Potato. The result of 20 years' investigation and experiment in cultivation and cooking. By McLarin. Notes by R. T. Trall, M.D. 50 cents.

Thoughts for the Young Men and Young Women of America. By L. U. Reavis. Notes by Greeley. $1.00.

Ways of Life, Showing the Right Way and the Wrong Way. Weaver. $1.

Weaver's Works. "Hopes and Helps," "Aims and Aids," "Ways of Life." A great work, in one vol. $3.00.

Wedlock. Right Relations of the Sexes. Laws of Conjugal Selection. Who may and who may not Marry. For both Sexes. By Wells. Plain, $1.50; gilt, 2$.

Capital Punishment; or, the Proper Treatment of Criminals. 10 cents.

Education of the Heart. Colfax. 10 cents. **Father Matthew,** Temperance Apostle, Portrait, Character and Biography. 10 cents.

History of Salem Witchcraft; The Planchette Mystery; Modern Spiritualism, By Mrs. Stowe; and Dr Doddridge's Celebrated Dream. 1 vol. $1.

AGENTS WANTED. There are many individuals in every neighborhood who would be glad to have one or more of these useful Books. HUNDREDS of COPIES could be sold *where they have never yet been introduced.* A *Local Agent* is wanted in every town, to whom liberal terms will be given. This is one way to do good and be paid for it. Send stamp for Illustrated Catalogue, with terms to Agents.

Address, S. R. WELLS & CO., 737 Broadway, New York.

www.ingramcontent.com/pod-product-compliance
Lightning Source LLC
Chambersburg PA
CBHW031354160426
43196CB00007B/807